Harcourt Math

Practice Workbook

TEACHER'S EDITION
Grade 6

Orlando • Boston • Dallas • Chicago • San Diego
www.harcourtschool.com

Copyright © by Harcourt, Inc.

All rights reserved. No part of this publication may be reproduced or transmitted in any form or by any means, electronic or mechanical, including photocopy, recording, or any information storage and retrieval system, without permission in writing from the publisher.

Permission is hereby granted to individual teachers using the corresponding student's textbook or kit as the major vehicle for regular classroom instruction to photocopy complete student pages from this publication in classroom quantities for instructional use and not for resale.

Duplication of this work other than by individual classroom teachers under the conditions specified above requires a license. To order a license to duplicate this work in greater than classroom quantities, contact Customer Service, Harcourt, Inc., 6277 Sea Harbor Drive, Orlando, Florida 32887-6777. Telephone: 1-800-225-5425. Fax: 1-800-874-6418 or 407-352-3445.

HARCOURT and the Harcourt Logo are trademarks of Harcourt, Inc.

Printed in the United States of America

ISBN 0-15-320798-1

4 5 6 7 8 9 10   073   10 09 08 07 06 05 04 03 02

CONTENTS

Unit 1: NUMBER SENSE AND OPERATIONS

Chapter 1: Whole Number Applications
- 1.1 Estimate with Whole Numbers 1
- 1.2 Use Addition and Subtraction 2
- 1.3 Use Multiplication and Division ... 3
- 1.4 Problem Solving Strategy: Predict and Test 4
- 1.5 Algebra: Use Expressions 5
- 1.6 Algebra: Mental Math and Equations 6

Chapter 2: Operation Sense
- 2.1 Mental Math: Use the Properties .. 7
- 2.2 Algebra: Exponents 8
- 2.4 Algebra: Order of Operations 9
- 2.5 Problem Solving Skill: Sequence and Prioritize Information 10

Chapter 3: Decimal Concepts
- 3.1 Represent, Compare, and Order Decimals 11
- 3.2 Problem Solving Strategy: Make a Table 12
- 3.3 Estimate with Decimals 13
- 3.4 Decimals and Percents 14

Chapter 4: Decimal Operations
- 4.1 Add and Subtract Decimals 15
- 4.2 Multiply Decimals 16
- 4.4 Divide with Decimals 17
- 4.5 Problem Solving Skill: Interpret the Remainder 18
- 4.6 Algebra: Decimal Expressions and Equations 19

Unit 2: STATISTICS AND GRAPHING

Chapter 5: Collect and Organize Data
- 5.1 Samples 20
- 5.2 Bias in Surveys 21
- 5.3 Problem Solving Strategy: Make a Table 22
- 5.4 Frequency Tables and Line Plots 23
- 5.5 Measures of Central Tendency 24
- 5.6 Outliers and Additional Data 25
- 5.7 Data and Conclusions 26

Chapter 6: Graph Data
- 6.1 Make and Analyze Graphs 27
- 6.2 Find Unknown Values............ 28
- 6.3 Stem-and-Leaf Plots and Histograms 29
- 6.5 Box-and-Whisker Graphs 30
- 6.6 Analyze Graphs 31

Unit 3: FRACTION CONCEPTS AND OPERATIONS

Chapter 7: Number Theory
- 7.1 Divisibility 32
- 7.2 Prime Factorization.............. 33
- 7.3 Least Common Multiple and Greatest Common Factor 34
- 7.4 Problem Solving Strategy: Make an Organized List 35

Chapter 8: Fraction Concepts
- 8.1 Equivalent Fractions and Simplest Form 36
- 8.2 Mixed Numbers and Fractions 37

- 8.3 Compare and Order Fractions **38**
- 8.5 Fractions, Decimals, and Percents . **39**

▶ **Chapter 9: Add and Subtract Fractions and Mixed Numbers**
- 9.1 Estimate Sums and Differences ... **40**
- 9.3 Add and Subtract Fractions **41**
- 9.4 Add and Subtract Mixed Numbers . **42**
- 9.6 Subtract Mixed Numbers **43**
- 9.7 Problem Solving Strategy: Draw a Diagram **44**

▶ **Chapter 10: Multiply and Divide Fractions and Mixed Numbers**
- 10.1 Estimate Products and Quotients **45**
- 10.2 Multiply Fractions **46**
- 10.3 Multiply Mixed Numbers **47**
- 10.5 Divide Fractions and Mixed Numbers **48**
- 10.6 Problem Solving Skill: Choose the Operation **49**
- 10.7 Algebra: Fraction Expressions and Equations **50**

▶ **Unit 4: ALGEBRA: INTEGERS**

▶ **Chapter 11: Number Relationships**
- 11.1 Understand Integers **51**
- 11.2 Rational Numbers **52**
- 11.3 Compare and Order Rational Numbers **53**
- 11.4 Problem Solving Strategy: Use Logical Reasoning **54**

▶ **Chapter 12: Operations with Integers**
- 12.2 Add Integers **55**
- 12.4 Subtract Integers **56**
- 12.5 Multiply and Divide Integers **57**
- 12.6 Explore Operations with Rational Numbers **58**

▶ **Unit 5: ALGEBRA: EXPRESSIONS AND EQUATIONS**

▶ **Chapter 13: Expressions**
- 13.1 Write Expressions **59**
- 13.2 Evaluate Expressions **60**
- 13.4 Expressions with Squares and Square Roots **61**

▶ **Chapter 14: Addition and Subtraction Equations**
- 14.1 Connect Words and Equations ... **62**
- 14.3 Solve Addition Equations **63**
- 14.4 Solve Subtraction Equations **64**

▶ **Chapter 15: Multiplication and Division Equations**
- 15.2 Solve Multiplication and Division Equations **65**
- 15.3 Use Formulas **66**
- 15.5 Problem Solving Strategy: Work Backward **67**

▶ **Unit 6: GEOMETRY AND SPATIAL REASONING**

▶ **Chapter 16: Geometric Figures**
- 16.1 Points, Lines, and Planes **68**
- 16.3 Angle Relationships **69**
- 16.4 Classify Lines **70**

▶ **Chapter 17: Plane Figures**
- 17.1 Triangles **71**
- 17.2 Problem Solving Strategy: Find a Pattern **72**
- 17.3 Quadrilaterals **73**
- 17.4 Draw Two-Dimensional Figures ... **74**
- 17.5 Circles **75**

Chapter 18: Solid Figures
- 18.1 Types of Solid Figures 76
- 18.2 Different Views of Solid Figures 77
- 18.4 Problem Solving Strategy: Solve a Simpler Problem 78

Chapter 19: Congruence and Similarity
- 19.1 Construct Congruent Segments and Angles 79
- 19.2 Bisect Line Segments and Angles 80
- 19.4 Similar and Congruent Figures 81

Unit 7: RATIO, PROPORTION, PERCENT, AND PROBABILITY

Chapter 20: Ratio and Proportion
- 20.1 Ratios and Rates 82
- 20.3 Problem Solving Strategy: Write an Equation 83
- 20.4 Algebra: Ratios and Similar Figures 84
- 20.5 Algebra: Proportions and Similar Figures 85
- 20.6 Algebra: Scale Drawings 86
- 20.7 Algebra: Maps 87

Chapter 21: Percent and Change
- 21.1 Percent 88
- 21.2 Percents, Decimals, and Fractions 89
- 21.3 Estimate and Find Percent of a Number 90
- 21.5 Discount and Sales Tax 91
- 21.6 Simple Interest 92

Chapter 22: Probability of Simple Events
- 22.1 Theoretical Probability 93
- 22.2 Problem Solving Skill: Too Much or Too Little Information 94
- 22.4 Experimental Probability 95

Chapter 23: Probability of Compound Events
- 23.1 Problem Solving Strategy: Make an Organized List 96
- 23.2 Compound Events 97
- 23.3 Independent and Dependent Events 98
- 23.4 Make Predictions 99

Unit 8: MEASUREMENT

Chapter 24: Units of Measure
- 24.1 Algebra: Customary Measurements 100
- 24.2 Algebra: Metric Measurements 101
- 24.3 Relate Customary and Metric 102
- 24.4 Appropriate Tools and Units 103
- 24.5 Problem Solving Skill: Estimate or Find Exact Answer 104

Chapter 25: Length and Perimeter
- 25.2 Perimeter 105
- 25.3 Problem Solving Strategy: Draw a Diagram 106
- 25.4 Circumference 107

Chapter 26: Area
- 26.1 Estimate and Find Area 108
- 26.2 Algebra: Areas of Parallelograms and Trapezoids 109

26.4 Algebra: Areas of Circles **110**
26.5 Algebra: Surface Areas of
     Prisms and Pyramids **111**

▶ **Chapter 27: Volume**
27.1 Estimate and Find Volume **112**
27.2 Problem Solving Strategy:
     Make a Model **113**
27.3 Algebra: Volumes of
     Pyramids **114**
27.5 Volumes of Cylinders **115**

▶ **Unit 9: ALGEBRA: PATTERNS AND RELATIONSHIPS**

▶ **Chapter 28: Algebra: Patterns**
28.1 Problem Solving Strategy:
     Find a Pattern **116**
28.2 Patterns in Sequences **117**
28.3 Number Patterns and
     Functions **118**
28.4 Geometric Patterns **119**

▶ **Chapter 29: Geometry and Motion**
29.1 Transformations of
     Plane Figures **120**
29.2 Tessellations **121**
29.3 Problem Solving Strategy:
     Make a Model **122**
29.4 Transformations of
     Solid Figures **123**
29.5 Symmetry **124**

▶ **Chapter 30: Algebra: Graph Relationships**
30.1 Inequalities on a
     Number Line **125**
30.2 Graph on the Coordinate
     Plane **126**
30.3 Graph Relations **127**
30.4 Problem Solving Skill: Make
     Generalizations **128**
30.6 Graph Transformations **129**

Name _____

LESSON 1.1

Estimate with Whole Numbers

Vocabulary

1. When both factors in a multiplication problem are rounded up to estimate the product, the estimate is an _____overestimate_____.

2. When all addends are about the same, you can use _____clustering_____ to estimate their sum.

Estimate the sum or difference. Possible estimates are given.

3. 2,489
 1,601
 +2,109
 6,000

4. 398
 415
 +368
 1,200

5. 4,723
 +2,198
 6,900

6. 7,132
 6,594
 +7,301
 21,000

7. 5,401
 +9,188
 14,600

8. 478
 − 26
 450

9. 263
 −211
 50

10. 5,877
 −5,318
 600

11. 8,528
 −6,491
 2,000

12. 8,903
 −4,575
 4,300

Estimate the product or quotient. Possible estimates are given.

13. 53
 × 8
 400

14. 76
 × 9
 720

15. 72
 ×28
 2,100

16. 47
 ×53
 2,500

17. 660
 × 42
 28,000

18. 371
 × 78
 32,000

19. 68
 ×37
 2,800

20. 480
 ×192
 100,000

21. 375
 ×591
 240,000

22. 824
 ×693
 560,000

23. 331 ÷ 5
 70

24. 643 ÷ 9
 70

25. 1,827 ÷ 59
 30

26. 5,543 ÷ 77
 70

27. 9,165 ÷ 28
 300

28. 6,281 ÷ 875
 7

29. 7,118 ÷ 614
 12

30. 8,215 ÷ 897
 9

Mixed Review

Round to the nearest 1,000.

31. 4,571
 5,000

32. 8,445
 8,000

33. 1,902
 2,000

34. 6,679
 7,000

Find the product.

35. 6 × 6 × 6
 216

36. 3 × 3 × 3 × 3
 81

37. 4 × 4 × 4 × 4
 256

Practice PW1

Name _____

LESSON 1.2

Use Addition and Subtraction

Find the sum or difference. Estimate to check. Possible estimates are given.

1. 504 + 343   2. 684 + 193 + 217   3. 3,991 − 953   4. 4,616 + 1,382

   800; 847   1,100; 1,094   3,000; 3,038   6,000; 5,998

5. 4,183 − 2,851   6. 794 + 578 + 909   7. 17,079 − 8,805   8. 29,114 − 13,513

   1,000; 1,332   2,300; 2,281   8,000; 8,274   15,000; 15,601

9. 12,379   10. 53,852   11. 60,118   12. 43,192   13. 72,583
   + 7,166   + 15,098   − 38,541   − 10,476   + 16,205
   19,000;   69,000;   20,000   30,000;   90,000;
   19,545   68,950   21,577   32,716   88,788

14. 68,450   15. 154,022   16. 864,191   17. 571,042   18. 389,077
    − 31,754   − 46,389   − 95,361   − 462,790   + 605,213
    40,000;   100,000;   800,000;   100,000;   1,000,000
    36,696   107,633   768,830   108,252   944,290

Solve.

19. 158 + 2,876 − 586   20. 1,422 + 806 + 539   21. 4,950 − 674 − 1,805

    2,448   2,767   2,471

22. 70,376 − 5,845 − 3,541   23. 8,026 + 11,061 + 3,824   24. 1,753 + 2,210 − 1,907

    60,990   22,911   2,056

25. 5,951 + 4,676 − 1,050 + 47,320   26. 19,321 − 1,322 + 939 − 3,084

    56,897   15,854

Mixed Review

Estimate the product or quotient. Possible estimates are given.

27.   57   28.   685   29.   173   30.   915   31.   718
    × 26   × 51   × 96   ×506   ×386
    1,800   35,000   20,000   450,000   280,000

32. 8,161 ÷ 87   33. 3,307 ÷ 47   34. 7,985 ÷ 432   35. 25,641 ÷ 197

    90   70   20   130

PW2 Practice

LESSON 1.3

Name _____

Use Multiplication and Division

Multiply or divide. Estimate to check.

1. 46 ×12	2. 230 × 15	3. 417 × 40	4. 2,515 × 52	5. 387 × 66
552	3,450	16,680	130,780	25,542
6. 217 ×154	7. 6,903 × 627	8. 582 ×316	9. 6,148 × 744	10. 8,132 × 915
33,418	4,328,181	183,912	4,574,112	7,440,780

11. $4\overline{)96}$ = 24
12. $9\overline{)423}$ = 47
13. $19\overline{)361}$ = 19
14. $7\overline{)756}$ = 108
15. $32\overline{)450}$ = 14 r2

16. $12\overline{)1,740}$ = 145
17. $19\overline{)912}$ = 48
18. $22\overline{)5,412}$ = 246
19. $31\overline{)4,836}$ = 156
20. $17\overline{)5,865}$ = 345

Divide. Write the remainder as a fraction.

21. $6\overline{)45}$ = $7\frac{1}{2}$
22. $14\overline{)550}$ = $39\frac{2}{7}$
23. $18\overline{)459}$ = $25\frac{1}{2}$
24. $41\overline{)13,210}$ = $322\frac{8}{41}$
25. $55\overline{)33,125}$ = $602\frac{3}{11}$

Mixed Review

Estimate the sum, difference, product, or quotient. Possible estimates are given.

26. 1,087
    2,109
  + 4,837
    8,000

27. 56,803
  − 31,942
    30,000

28. 347
    ×261
    90,000

29. 26,811 ÷ 885   30

Solve by using addition and subtraction.

30. 9,271 − 3,587 − 1,266 − 2,650 _____ 1,768

31. 2,114 + 739 + 4,799 + 557 + 1,632 _____ 9,841

Practice PW3

Name _____

LESSON 1.4

Problem-Solving Strategy: Predict and Test

Solve by predicting and testing.

1. Ryan bought a total of 40 juice boxes. He bought 8 more boxes of apple juice than of grape juice. How many of each kind did he buy?

   _____24 apple juice, 16 grape juice_____

2. The perimeter of a rectangular garden is 56 ft. The length is 4 ft more than the width. What are the dimensions of the garden?

   _____$l = 16$ ft; $w = 12$ ft_____

3. The Hawks soccer team played a total of 24 games. They won 6 more games than they lost, and they tied 2 games. How many games did they win?

   _____14 games_____

4. Rico collected a total of 47 rocks. He gathered 5 more jagged rocks than smooth rocks. How many of each kind of rock did he collect?

   _____26 jagged rocks, 21 smooth rocks_____

5. Matt has earned $75. To buy a bicycle, he needs twice that amount plus $30. How much does the bicycle cost?

   _____$180_____

6. The perimeter of a rectangular lot is 190 ft. The width of the lot is 15 ft more than the length. What are the dimensions of the lot?

   _____$w = 55$ ft; $l = 40$ ft_____

7. The Wolverines swimming team won a total of 15 first- and second-place medals at their last swim meet. If they won 7 more first-place medals than second-place medals, how many first-place medals did they win?

   _____11 first-place medals_____

8. Valley High School's football team played a total of 16 games. They won twice as many games as they lost. If they tied one game, how many games did the team win?

   _____10 games_____

Mixed Review

Find the product or quotient. Estimate to check. Possible estimates are given.

9. 306×582

   _____180,000; 178,092_____

10. $8,246 \div 38$

    _____200; 217_____

11. $21,420 \div 51$

    _____400; 420_____

Tell whether the estimate is an *overestimate* or *underestimate*. Then show how the estimate was determined.

12. $1,872 + 4,774 \approx 7,000$ _____overestimate; 2,000 + 5,000_____

13. $321 \times 82 \approx 24,000$ _____underestimate; 300 × 80_____

PW4 Practice

Name _____

LESSON 1.5

Algebra: Use Expressions

Vocabulary

Write the correct letter from Column 2.

Column 1

___a___ 1. a mathematical phrase that includes only numbers and operation symbols

___c___ 2. an expression that includes a variable

___b___ 3. a letter or symbol that stands for one or more numbers

Column 2

a. numerical expression

b. variable

c. algebraic expression

Write a numerical or algebraic expression for the word expression.

4. seven less than eleven

   _____$11 - 7$_____

5. six more than a number, x

   _____$x + 6$_____

6. 8 multiplied by m

   _____$m \times 8$_____

7. 84 divided by 8

   _____$84 \div 8$_____

Evaluate each expression.

8. 19×48

   _____912_____

9. $63b$, for $b = 15$

   _____945_____

10. $w + 178$, for $w = 226$

    _____404_____

11. $a \div b$, for $a = 253$ and $b = 11$

    _____23_____

12. $h + k - 84$, for $h = 46$ and $k = 73$

    _____35_____

13. $r(s)$, for $r = 109$ and $s = 33$

    _____3,597_____

Mixed Review

Multiply or divide.

14. $18\overline{)1{,}854}$ → 103

15. 631×55 = 34,705

16. 490×117 = 57,330

17. $54\overline{)11{,}988}$ → 222

18. Use the table at the right. If the pattern continues, how many laps in all will 8 swimmers swim on the fourth day?

    _____96 laps_____

Each Swimmer's Training Schedule				
Day	1	2	3	4
Laps	6	8	10	☐

Practice PW5

Name _____

LESSON 1.6

Algebra: Mental Math and Equations

Determine which of the given values is a solution of the equation.

1. $4d = 28$;
   $d = 7, 8,$ or 9

   _____ $d = 7$ _____

2. $50 - t = 28$;
   $t = 20, 21,$ or 22

   _____ $t = 22$ _____

3. $42 \div n = 6$;
   $n = 5, 6,$ or 7

   _____ $n = 7$ _____

4. $72 + v = 85$;
   $v = 12, 13,$ or 14

   _____ $v = 13$ _____

5. $m + 7 = 18$;
   $m = 9, 10,$ or 11

   _____ $m = 11$ _____

6. $s - 17 = 10$;
   $s = 26, 27,$ or 28

   _____ $s = 27$ _____

7. $c \div 8 = 3$;
   $c = 22, 23,$ or 24

   _____ $c = 24$ _____

8. $155 = 5k$;
   $k = 30, 31,$ or 32

   _____ $k = 31$ _____

9. $8 = 25 - x$;
   $x = 17, 18,$ or 19

   _____ $x = 17$ _____

Solve each equation by using mental math.

10. $e + 6 = 20$

    _____ $e = 14$ _____

11. $x \div 2 = 10$

    _____ $x = 20$ _____

12. $6 \times h = 300$

    _____ $h = 50$ _____

13. $s - 18 = 40$

    _____ $s = 58$ _____

14. $92 = b + 7$

    _____ $b = 85$ _____

15. $90 \div t = 15$

    _____ $t = 6$ _____

16. $m - 150 = 420$

    _____ $m = 570$ _____

17. $8 \times n = 72$

    _____ $n = 9$ _____

18. $f - 6 = 98$

    _____ $f = 104$ _____

19. $c \times 4 = 40$

    _____ $c = 10$ _____

20. $63 = d \times 7$

    _____ $d = 9$ _____

21. $k + 28 = 32$

    _____ $k = 4$ _____

22. $9x = 180$

    _____ $x = 20$ _____

23. $6 = v - 58$

    _____ $v = 64$ _____

24. $w \div 9 = 12$

    _____ $w = 108$ _____

25. $p + 62 = 100$

    _____ $p = 38$ _____

Mixed Review

Find the sum or difference. Estimate to check. Possible estimates are given.

26. 390
     $+789$
     $1,200;$
     $1,179$

27. $9,056$
     $-1,732$
     $7,000;$
     $7,324$

28. $1,978$
     $+693$
     $2,700;$
     $2,671$

29. $47,813$
     $-9,507$
     $40,000;$
     $38,306$

30. $73,681$
     $+50,342$
     $120,000;$
     $124,023$

Evaluate each expression.

31. $n + 701$, for $n = 510$

    _____ $1,211$ _____

32. $50p$, for $p = 53$

    _____ $2,650$ _____

33. $r \times s$, for $r = 12$ and $s = 30$

    _____ 360 _____

34. $h + g$, for $h = 65$ and $g = 41$

    _____ 106 _____

Name _____

LESSON 2.1

Use the Properties

Vocabulary

Write the correct letter from Column 2.

Column 1   **Column 2**

__b__ 1. Associative Property   a. $58 + 72 = (58 + 2) + (72 - 2)$

__c__ 2. Commutative Property   b. $3 \times (2 \times 4) = (3 \times 2) \times 4$

__a__ 3. compensation   c. $10 \times 23 = 23 \times 10$

__e__ 4. Distributive Property   d. $18x = 18$

__d__ 5. Identity Property of One e. $6 \times 24 = 6 \times (20 + 4)$

Use mental math to find the value.

6. $37 + 14$ __51__

7. $65 - 23$ __42__

8. 18×6 __108__

9. $258 \div 3$ __86__

10. 18×22 __396__

11. $141 \div 3$ __47__

12. $78 - 45$ __33__

13. $49 + 14$ __63__

14. $41 + 18$ __59__

15. 19×11 __209__

16. $37 - 11$ __26__

17. $366 \div 6$ __61__

18. $320 \div 5$ __64__

19. $59 + 26$ __85__

20. $74 - 23$ __51__

21. 15×51 __765__

22. $88 - 54$ __34__

23. 43×21 __903__

24. $465 \div 15$ __31__

25. $56 + 15$ __71__

26. 15×48 __720__

27. $32 + 35$ __67__

28. $153 \div 9$ __17__

29. $96 - 25$ __71__

30. $37 + 14 + 43$ __94__

31. $(7 \times 12) + (7 \times 18)$ __210__

32. $5 \times 33 \times 6$ __990__

Mixed Review

Evaluate each expression for $a = 72$, $b = 28$, and $c = 8$.

33. $b \times 7$

34. $a + b + 362$

35. $a \div c$

36. $225 - a$

__196__

__462__

__9__

__153__

Solve each equation using mental math.

37. $n \times 8 = 56$

38. $19 + w = 36$

39. $h \div 20 = 35$

40. $98 - x = 59$

__n = 7__

__w = 17__

__h = 700__

__x = 39__

Practice PW7

Name _____

LESSON 2.2

Exponents

Vocabulary

Complete using *exponent* or *base*.

1. A(n) ____exponent____ shows how many times a number called the ____base____ is used as a factor.

Write the equal factors. Then find the value.

2. 5^4 3. 10^5 4. 18^2

   $5 \times 5 \times 5 \times 5 = 625$ $10 \times 10 \times 10 \times 10 \times 10 = 100{,}000$ $18 \times 18 = 324$

5. 2^6 6. 15^1 7. 4^3

   $2 \times 2 \times 2 \times 2 \times 2 \times 2 = 64$ 15 $4 \times 4 \times 4 = 64$

Write in exponent form.

8. $1 \times 1 \times 1$ 9. $n \times n \times n \times n$ 10. $6 \times 6 \times 6 \times 6 \times 6$

   1^3 n^4 6^5

11. $10 \times 10 \times 10 \times 10$ 12. $y \times y$ 13. $4 \times 4 \times 4 \times 4 \times 4 \times 4$

   10^4 y^2 4^6

Express with an exponent and the given base.

14. 125, base 5   15. 256, base 4   16. 729, base 9

   5^3 4^4 9^3

17. 64, base 2   18. 81, base 3   19. 1,000,000, base 10

   2^6 3^4 10^6

Mixed Review

Use mental math to find the value.

20. $65 + 27$ 21. $20 \times 14 \times 5$ 22. $(9 \times 4) + (9 \times 6)$

   92 $1{,}400$ 90

23. $84 - 45$ 24. $3 \times 3 \times 3 \times 3$ 25. 7^2

   39 81 49

PW8   Practice

Name _____

LESSON 2.4

Order of Operations

Give the correct order of operations.

1. $100 + 6^2 - 9$

    _____Clear exponents._____

    _____Add. Subtract._____

2. $(52 - 49)^2 \div 9$

    _____Operate inside parentheses._____

    _____Clear exponents. Divide._____

3. $(5^2 + 1) \div 2$

    _____Inside parentheses: Clear_____

    _____exponents and add. Divide._____

4. $(9 + 2) \times (16 - 12)^2$

    _____Operate inside parentheses._____

    _____Clear exponents. Multiply._____

Evaluate the expression.

5. $27 \div 3 + 1$

    _____10_____

6. $(6 + 8) \times (9 - 8)$

    _____14_____

7. $(6 + 7^2) \div 5 \times 2$

    _____22_____

8. $(12 \div 2)^3 + (2^3 + 1^3)$

    _____225_____

9. $(15 - 5)^2 - (4 \times 3)$

    _____88_____

10. $(57 + 3) \times 2^4$

    _____960_____

11. $(19 + 9) \div (2^3 - 1) + 20$

    _____24_____

12. $(3 \times 7^2) - (5^3 - 9^2) + 10^2$

    _____203_____

13. $3 \times (10^2 - 65) + (5^2 \times 2)$

    _____155_____

Evaluate the expression for $s = 5$ and $t = 12$.

14. $50 \div s + 7$

    _____17_____

15. $s^2 + 150$

    _____175_____

16. $2 \times t - 18$

    _____6_____

17. $t^2 - 3 \times 8$

    _____120_____

18. $15 + t \div 6$

    _____17_____

19. $27 + 9 \times s$

    _____72_____

Mixed Review

Use mental math to find the value.

20. 12×7

    _____84_____

21. $37 + 62$

    _____99_____

22. $434 \div 7$

    _____62_____

23. $1{,}731 - 605$

    _____1,126_____

Write in exponent form.

24. $8 \times 8 \times 8 \times 8$

    _____8^4_____

25. $6 \times 6 \times 6 \times 6 \times 6$

    _____6^5_____

26. $n \times n \times n \times n \times n$

    _____n^5_____

Practice PW9

Name _____

LESSON 2.5

Problem Solving Skill

Sequence and Prioritize Information

Tiffany and her dad need to make brownies for the PTA bake sale. They need to deliver the brownies to the school by 1 P.M. To plan their morning, they made a list of the things they need to do, including a time estimate for each task.

To Do List
- Bake brownies, 20 minutes.
- Let brownies cool, 20 minutes.
- Grocery shopping, buy brownie ingredients, 1 hour.
- Drive to the school, 10 minutes.
- Mix brownies, 30 minutes.
- Wrap brownies separately in plastic wrap, 15 minutes.

1. List the items in the To Do List in an order that makes sense.

   grocery shopping, mix brownies

   bake brownies, cool brownies,

   wrap brownies, drive to school

2. Can they get everything done if they begin at noon? Explain.

   No. It takes 2 hr 35 min to complete all of the tasks, so they

   would not get the brownies to the school until 2:35 P.M.

Alex has several things to do on Saturday.

Saturday Activities
- Attend birthday party at 4 P.M.
- Buy gift—either a CD for $16 or a computer game for $25.
- Get haircut at 2 P.M.; cost $9.
- Before 10 A.M., mow Mrs Brown's lawn; earn $15.
- Mow Mr. Tanaka's lawn after 10:00 A.M. earn $15. Trim hedge, earn $10.
- Set aside $5 for savings.
- Keep $5 for spending money.

3. List his activities in an order that makes sense.

   Possible answer: mow Mrs. Brown's

   lawn, mow Mr. Tanaka's lawn/trim

   hedge, get haircut, set aside $5 for

   savings, keep $5 for spending money,

   buy gift, attend birthday party.

4. Which gift can Alex buy? Why? Assume he has no spending money left from last week.

   The CD; because he has only $21 to spend on the gift.

Mixed Review

Evaluate each expression.

5. $t \times 7$, for $t = 25$ 6. $150 - h$, for $h = 88$ 7. $96 \div r$, for $r = 2$

   175   62   48

PW10 Practice

LESSON 3.1

Name _____

Represent, Compare, and Order Decimals

Write the value of the underlined digit.

1. 485.03<u>6</u>
   6 thousandths

2. 16,005.8<u>4</u>5
   4 hundredths

3. 8,492.<u>7</u>792
   7 tenths

Write the number in expanded form.

4. 5.71 $5 + 0.7 + 0.01$

5. 85.083 $80 + 5 + 0.08 + 0.003$

6. 0.4625 $0.4 + 0.06 + 0.002 + 0.0005$

7. 17.00157 $10 + 7 + 0.001 + 0.0005 + 0.00007$

Compare the numbers. Write $<$, $>$, or $=$ for ◯.

8. 15.4 ◯ 14.5 $>$

9. 5.67 ◯ 5.76 $<$

10. 43.90 ◯ 43.9 $=$

11. 7.91 ◯ 9.17 $<$

12. 765.28 ◯ 762.58 $>$

13. 0.234 ◯ 2.304 $<$

Write the numbers in order from least to greatest.

14. 3,224; 2,432; 3,422
    2,432; 3,224; 3,422

15. 88.5; 85.8; 58.8
    58.8; 85.8; 88.5

16. 6.21; 6.02; 6.12
    6.02; 6.12; 6.21

Write the numbers in order from greatest to least

17. 0.005; 0.500; 0.050
    0.500; 0.050; 0.005

18. 317.8; 318.7; 371.8
    371.8; 318.7; 317.8

19. 16.04; 14.6; 16.4
    16.4; 16.04; 14.6

Mixed Review

Evaluate each expression.

20. $4 + 3^3 \times 2 - (6 - 1)$
    53

21. $(11 + 16) \div 3 + (4 - 2)^2$
    13

22. $45 + (6^2 - 11) \times 2$
    95

Solve each equation by using mental math.

23. $m - 7 = 36$
    $m = 43$

24. $9x = 63$
    $x = 7$

25. $a \div 6 = 14$
    $a = 84$

Evaluate each expression for $a = 6$, $b = 120$, and $c = 54$.

26. $b + 295$
    415

27. $93 - c$
    39

28. $b \div a$
    20

Practice   **PW11**

Name _____   **LESSON 3.2**

Problem Solving Strategy: Make a Table

Solve the problem by making a table.

1. Earthquakes are measured using the Richter scale. The greater the number, the greater the magnitude (or strength). Some of the strongest earthquakes during the twentieth century had magnitudes of 7.2, 8.9, 8.4, 8.7, 8.3, 8.6, 7.7, and 8.1. The San Francisco earthquake of 1906 had the fifth highest magnitude of those given above. What was its magnitude on the Richter scale?

   _____8.3_____

2. Late in 1999, one U.S. dollar was worth the following amounts in five other countries' money.

Australian dollar	1.5798
Brazilian real	1.8780
Canadian dollar	1.4796
German mark	1.9524
Swiss franc	1.5919

   In which country could one U.S. dollar be exchanged for the greatest amount of that country's money?

   _____Germany_____

3. Danny is doing library research on animals. He has spent 25 minutes reading about insects. He thinks he will need the same amount of time for each of 5 other types of animals. If he began at 9:45 A.M., at what time would he finish?

   _____12:15 P.M._____

4. A theater is showing two films. The starting times for the first film are every even hour, beginning at noon. The starting times for the second film are every odd hour, beginning at 1:00 P.M. If the last show begins at 10:00 P.M., how many times are both films shown?

   _____11 times (1st: 6; 2nd: 5)_____

Use the table at the right for 5 and 6. The numbers are amounts of energy in quadrillion BTUs.

5. In which country is the difference between amount of energy produced and amount used the greatest?

   _____United States_____

Country	Energy Produced	Energy Used
United States	66.68	82.19
Great Britain	9.23	9.68
China	30.18	29.22
Canada	14.36	10.97
India	6.94	8.51
Russia	45.66	32.72

6. In which country is the difference between amount of energy produced and amount used the least? _____Great Britain_____

Mixed Review

Use mental math to find the value.

7. 67 + 83 + 33

   _____183_____

8. 449 − 398

   _____51_____

9. 203 + 178 + 22

   _____403_____

Write which operation you would do first.

10. 8 − 5 + 7

    _____subtraction_____

11. 16 + 4 ÷ 2

    _____division_____

12. (10 + 4) × 2

    _____addition_____

PW12 Practice

Estimate with Decimals

Estimate. Possible estimates are given.

1. $3.8 + 7.9$
   _12_

2. 7.1×6.2
   _42_

3. $23.18 - 19.09$
   _4_

4. $12.2 \div 5.9$
   _2_

5. 4.09×6.18
   _24_

6. $83.89 + 17.66$
   _102_

7. $162.3 \div 15.7$
   _10_

8. $31.6 - 8.82$
   _23_

9. $7.7 + 118.2$
   _126_

10. $101.2 - 34.9$
    _66_

11. $35.99 - 6.02
    _$30_

12. 19.8×21.3
    _400_

13. 124.66×3
    _$375_

14. $10.6 + 19.01$
    _30_

15. 81.3×9.6
    _800_

16. $810.1 - 69.9$
    _740_

17. $602.5 + 87.3$
    _690_

18. 397.9×21
    _8,000_

19. $502.03 \div 4.9$
    _100_

20. $88.20 + 79.10
    _$170_

21. $1.8 + 2.9 + 11.8$
    _17_

22. $203.99 \div 21$
    _$10_

23. $199.50 - 53.99
    _$145_

24. 8.8×7.1
    _63_

25. $67.2 + 11.9 + 107.44$
    _190_

26. $889.52 - 402.68$
    _490_

Mixed Review

Write in exponent form.

27. $4 \times 4 \times 4$
    _4^3_

28. $2 \times 2 \times 2 \times 2$
    _2^4_

29. 6×6
    _6^2_

30. $1 \times 1 \times 1 \times 1 \times 1$
    _1^5_

31. $7 \times 7 \times 7 \times 7$
    _7^4_

32. $8 \times 8 \times 8$
    _8^3_

33. $9 \times 9 \times 9$
    _9^3_

34. $3 \times 3 \times 3 \times 3$
    _3^4_

Find the value.

35. 5^2
    _25_

36. 2^5
    _32_

37. 8^2
    _64_

38. 1^4
    _1_

Name _____

LESSON 3.4

Decimals and Percents

Write the decimal and percent for the shaded part.

1.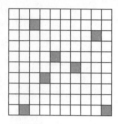
2.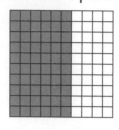
3.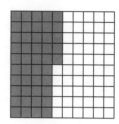

   _0.07, 7%_   _0.60, or 0.6, 60%_   _0.45, 45%_

4.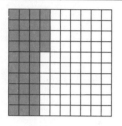
5.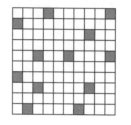
6.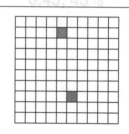

   _0.34, 34%_   _0.12, 12%_   _0.02, 2%_

Write the corresponding decimal or percent.

7. 67%   8. 0.15   9. 0.92   10. 11%   11. 80%
   _0.67_   _15%_   _92%_   _0.11_   _0.80 or 0.8_

12. 0.3   13. 64%   14. 88%   15. 0.14   16. 90%
    _30%_   _0.64_   _0.88_   _14%_   _0.90 or 0.9_

17. 0.09   18. 34%   19. 0.75   20. 6%   21. 0.19
    _9%_   _0.34_   _75%_   _0.06_   _19%_

Mixed Review

Evaluate the expression.

22. $6 + 3 \times 2$ 23. $10 \div 2 - 1$ 24. $16 - 4 \times 2$
    _12_   _4_   _8_

25. $20 \times 2 + 1$ 26. $(8 - 2) \times 3$ 27. $15 \div 3 + 2$
    _41_   _18_   _7_

28. $32 + 8 \div 2$ 29. $20 + (6 \times 2)$ 30. $45 \div (4 + 5)$
    _36_   _32_   _5_

31. $(16 - 7)^2 \div 3$ 32. $6^2 + 14 \div 2$ 33. $12 \times (9 - 4)$
    _27_   _43_   _60_

PW14 Practice

Name _____

LESSON 4.1

Add and Subtract Decimals

Add or subtract. Estimate to check.

1. 0.34 + 8.19
   _8.53_
2. 6.92 + 3.55
   _10.47_
3. 0.418 + 1.291
   _1.709_
4. 8.93 + 2.68
   _11.61_

5. 8.7 − 4.2
   _4.5_
6. 13.29 − 5.96
   _7.33_
7. 5.41 − 1.36
   _4.05_
8. 15.93 − 7.08
   _8.85_

9. 9.328 + 1.294
   _10.622_
10. 5.962 − 1.748
    _4.214_
11. 4.036 − 2.751
    _1.285_
12. 4.89 + 12.45
    _17.34_

13. 8.116 − 3.094
    _5.022_
14. 23.4 − 12.379
    _11.021_
15. 20.68 + 7.12
    _27.8_
16. 1.681 + 2.899
    _4.58_

17. 41.783 − 29.822
    _11.961_
18. 21.35 + 37.7 + 12.816
    _71.866_
19. $245.62 − $109.99
    _$135.63_

20. 41.6 + 27.56 + 16.942
    _86.102_
21. 452.803 − 376.991
    _75.812_
22. 111.22 + 77.5 + 83.947
    _272.667_

23. 446.09
     811.36
    + 73.52
    _1,330.97_

24.   8.71
     13.99
    + 67.2
    _89.9_

25. 89.01
    − 67.56
    _21.45_

26. 25.8
    − 17.226
    _8.574_

27.   23.75
      873.33
    + 2,586.02
    _3,483.10_

28. 71.043
    − 58.649
    _12.394_

29. 36.583
    − 16.007
    _20.576_

30.   3.056
    1,691.396
    + 44.21
    _1,738.662_

Mixed Review

Write the numbers in order from greatest to least.

31. 21.10; 21.050; 21.8
    _21.8; 21.10; 21.050_
32. 36.63; 36.33; 36.36
    _36.63; 36.36; 36.33_
33. 5.912; 5.921; 5.192
    _5.921; 5.912; 5.192_

Write the percent or decimal.

34. 0.98
    _98%_
35. 73%
    _0.73_
36. 44%
    _0.44_
37. 0.06
    _6%_
38. 90%
    _0.90 or 0.9_

Practice **PW15**

Name _____

LESSON 4.2

Multiply Decimals

Tell the number of decimal places there will be in the product.

1. 6.3×0.75 2. 9.7×48.8 3. 5.96×62.15 4. 37.6×8.3

   ___3___   ___2___   ___4___   ___2___

5. 32.08×7.3 6. 428.9×5.6 7. 897.3×5.3 8. 186.472×9.6

   ___3___   ___2___   ___2___   ___4___

Place the decimal point in the product.

9. $6.17 \times 8.2 = 50594$ 10. $24.01 \times 8.51 = 2043251$ 11. $8.94 \times 5.27 = 471138$

   ___50.594___   ___204.3251___   ___47.1138___

12. $8.04 \times 1.7 = 13668$ 13. $19.6 \times 5.8 = 11368$ 14. $30.7 \times 8.33 = 255731$

    ___13.668___   ___113.68___   ___255.731___

Multiply. Estimate to check.

15. 5×0.9 16. 9×1.2 17. 4×3.47 18. $\$18.93 \times 7$

    ___4.5___   ___10.8___   ___13.88___   ___$132.51___

19. 5.55×9 20. 5×2.89 21. 31.82×4 22. 4.61×8

    ___49.95___   ___14.45___   ___127.28___   ___36.88___

23. 2.49×6 24. 35.98×6.3 25. 73.02×9.1 26. 8.5×16.03

    ___14.94___   ___226.674___   ___664.482___   ___136.255___

27. 3.91×6.22 28. 164.5×0.03 29. 28.14×1.52 30. 6.114×3.72

    ___24.3202___   ___4.935___   ___42.7728___   ___22.74408___

Mixed Review

Write the decimal and the percent for the shaded part.

31. 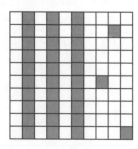   32. 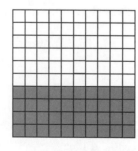   33.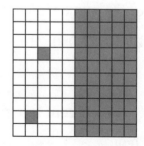

    ___0.33, 33%___   ___0.40, or 0.4, 40%___   ___0.52, 52%___

PW16 Practice

Name _____

LESSON 4.4

Divide with Decimals

Rewrite the problem so that the divisor is a whole number.

1. $8.5 \div 2.3$
2. $6.4 \div 1.3$
3. $9.1 \div 0.15$
4. $33.17 \div 6.8$

   $85 \div 23$
   $64 \div 13$
   $910 \div 15$
   $331.7 \div 68$

Place the decimal point in the quotient.

5. $7.48 \div 0.25 = 2992$
6. $116.13 \div 4.2 = 2765$
7. $56.68 \div 0.08 = 7085$

   29.92
   27.65
   708.5

Divide. Estimate to check.

8. $36.9 \div 3$
9. $22.4 \div 7$
10. $37.5 \div 5$
11. $89.6 \div 8$

    12.3
    3.2
    7.5
    11.2

12. $14)\overline{78.4}$ 5.6
13. $40)\overline{6.8}$ 0.17
14. $13)\overline{150.8}$ 11.6
15. $70)\overline{23.8}$ 0.34

16. $5.32 \div 0.7$
17. $1.88 \div 0.4$
18. $2.12 \div 0.2$
19. $5.4 \div 0.08$

    7.6
    4.7
    10.6
    67.5

20. $7.54)\overline{24.882}$ 3.3
21. $12.6)\overline{806.4}$ 64
22. $0.91)\overline{6.734}$ 7.4
23. $10.9)\overline{81.75}$ 7.5

24. $2.9)\overline{0.3335}$ 0.115
25. $0.18)\overline{64.296}$ 357.2
26. $12.3)\overline{84.87}$ 6.9
27. $8.7)\overline{53.244}$ 6.12

Mixed Review

Add, subtract, or multiply.

28. 78.94
     9.66
     $+ 103.71$
     192.31

29. $1,083.75$
     $- 706.9$
     376.85

30. 0.072
     $\times 0.48$
     0.03456

31. $215.6 + 49.87 + 8.351$
32. 42.83×1.91
33. $65.85 - 39.478$

    273.821
    81.8053
    26.372

34. $430.62 - 288.74$
35. $192.6 + 847.56$
36. 17.335×8.26

    141.88
    $1,040.16$
    143.1871

Practice **PW17**

Name _____   LESSON 4.5

Problem Solving Skill: Interpret the Remainder

Solve the problem by interpreting the remainder.

1. Thirty-seven people are attending a party at a restaurant. In the banquet room, the restaurant staff has set up tables that can each seat 8 people. What is the least number of tables that the group will use?

   _____5 tables_____

2. There are 23 pancakes on the griddle at a restaurant. The chef places 4 pancakes on each order. How many orders can the chef fill, and how many pancakes must be added to those remaining to make another order?

   _____5 orders; 1 pancake_____

3. A library reading room contains a number of tables that can seat 4 people. What is the least number of tables needed to seat 54 people?

   _____14 tables_____

4. A group of 5 friends wants to buy snacks. If each snack costs $0.75 and they have a total of $4.80 to spend, how many snacks can they buy?

   _____6 snacks_____

5. The chef at a restaurant uses 3 eggs to make each omelet. If the chef has 200 eggs, how many 3-egg omelets can he make?

   _____66 omelets_____

6. A total of 125 hamburgers were sold at a fund-raiser at the last football game. If the hamburger patties came in packages of 8, how many packages were opened?

   _____16 packages_____

Mixed Review

Estimate the sum or difference. Possible estimates are given.

7. $\begin{array}{r} 671 \\ +902 \\ \hline 1{,}600 \end{array}$

8. $\begin{array}{r} 478 \\ -310 \\ \hline 200 \end{array}$

9. $\begin{array}{r} 831 \\ -289 \\ \hline 500 \end{array}$

10. $\begin{array}{r} 1{,}226 \\ +533 \\ \hline 1{,}700 \end{array}$

11. $\begin{array}{r} 661 \\ +2{,}403 \\ \hline 3{,}100 \end{array}$

12. $\begin{array}{r} 1{,}729 \\ -494 \\ \hline 1{,}200 \end{array}$

13. $\begin{array}{r} 488 \\ -391 \\ \hline 100 \end{array}$

14. $\begin{array}{r} 2{,}994 \\ +1{,}258 \\ \hline 4{,}000 \end{array}$

Solve each equation by using mental math.

15. $m + 12 = 15$ 16. $5w = 20$ 17. $x - 7 = 8$ 18. $q + 4 = 10 + 6$

   $m = 3$ $w = 4$ $x = 15$ $q = 12$

19. $6r = 24$ 20. $y - 9 = 10$ 21. $a - 2 = 8 + 6$ 22. $d + 3 = 21 - 7$

   $r = 4$ $y = 19$ $a = 16$ $d = 11$

PW18   Practice

Name _____

LESSON 4.6

Algebra: Decimal Expressions and Equations

Evaluate each expression.

1. $t - 1.2$ for $t = 3$

   __1.8__

2. $y + 4.6$ for $y = 2.4$

   __7__

3. $8.2 - m$ for $m = 1.1$

   __7.1__

4. $2.4 \div a$ for $a = 6$

   __0.4__

5. $6g$ for $g = 1.5$

   __9__

6. $j - 6.3$ for $j = 9.6$

   __3.3__

7. $12.6 + r$ for $r = 4.4$

   __17__

8. $4.5 \div p$ for $p = 9$

   __0.5__

9. $7.24 - q$ for $q = 1.04$

   __6.20 or 6.2__

10. $6.18 \div y$ for $y = 3$

    __2.06__

11. $t + 4.66$ for $t = 2.1$

    __6.76__

12. $5h$ for h $= 2.4$

    __12__

Solve each equation by using mental math.

13. $w + 4.5 = 8$

    __$w = 3.5$__

14. $\frac{k}{3} = 2.5$

    __$k = 7.5$__

15. $1.4 = \frac{t}{2}$

    __$t = 2.8$__

16. $m - 7.6 = 2.4$

    __$m = 10$__

17. $3a = 6.9$

    __$a = 2.3$__

18. $9c = 22.5$

    __$c = 2.5$__

19. $3b = 6.4 + 2.6$

    __$b = 3$__

20. $w + 10.3 = 21.7$

    __$w = 11.4$__

21. $13.7 = d - 3.4$

    __$d = 17.1$__

22. $4.8 = \frac{n}{4}$

    __$n = 19.2$__

23. $\frac{x}{5} = 19.5$

    __$x = 97.5$__

24. $7h = 15.4$

    __$h = 2.2$__

Mixed Review

Estimate. Possible estimates are given.

25. $6.9 + 7.8$

    __15__

26. 31.77×6

    __180__

27. $63.85 \div 8$

    __8__

28. $17.04 - 9.8$

    __7__

29. $18.58 + 21.44$

    __40__

30. 91.92×4

    __360__

31. $54.3 - 19.7$

    __34__

32. $80.8 \div 9.2$

    __9__

Find the quotient.

33. $88.8 \div 6$

    __14.8__

34. $59.4 \div 36$

    __1.65__

35. $38.88 \div 7.2$

    __5.4__

36. $31.108 \div 2.2$

    __14.14__

Practice PW19

Name _____

LESSON 5.1

Samples

Determine the type of sample. Write *convenience, random,* or *systematic.*

1. An assembly-line worker randomly selected one microwave oven and then checked every fifteenth oven to see whether it worked.

   _____systematic_____

2. Carl selected students to complete a survey by assigning each student's name a number from 1 to 6, rolling a cube numbered 1 to 6, and choosing each student whose name had the number he rolled.

   _____random_____

3. A store manager asked the first 50 shoppers to enter her store on Saturday to complete a survey about changes they would like to see made at the store.

   _____convenience_____

Tell whether you would survey the population or use a sample. Explain.

4. You want to know the type of computer, if any, that each student in your class has at home.

   population; There are not too many class
   members to survey them all.

5. You want to know the average number of siblings of all sixth grade students in your school district.

   Possible answer: sample; There are too many
   sixth graders to survey them all.

6. You want to know your friends' favorite television program.

   Possible answer: population; Your group of
   friends is not too large to survey them all.

Mixed Review

Evaluate each expression.

7. $9.03 \div x$ for $x = 3$ 8. $7m$ for $m = 2.2$ 9. $4.5 - w$ for $w = 1.9$

   _____3.01_____   _____15.4_____   _____2.6_____

10. $17.4 + h$ for $h = 5.9$ 11. $k \div 2$ for $k = 6.4$ 12. $6.58 + a$ for $a = 0.45$

    _____23.3_____   _____3.2_____   _____7.03_____

PW20 Practice

Name _____

Bias in Surveys

Vocabulary

Complete.

1. A sample is ____biased____ if individuals in the population are not represented in the sample.

Tell whether the sampling method is *biased* or *unbiased*. Explain.

The Tri-State Soccer League is conducting a survey to determine if the players want to change the style of soccer shirt.

2. Randomly survey all players who wear size large shirts.

    __biased; excludes players__
    __who wear other sizes__

3. Randomly survey all members of championship teams.

    __biased; excludes members of__
    __non-championship teams__

4. Randomly survey 80 players.

    __unbiased; all players in the__
    __league have an equal chance of__
    __being selected__

5. Randomly survey all league coaches.

    __biased; excludes all players__

Determine whether the question is biased. Write *biased* or *unbiased*.

6. Do you feel that country music is better than all other types of music?

    __biased__

7. What type of team sport do you enjoy playing?

    __unbiased__

Mixed Review

Solve each equation by using mental math.

8. $w - 7.5 = 12.3$

    $w = 19.8$

9. $5x = 16.5$

    $x = 3.3$

10. $a + 6.9 = 14.3$

    $a = 7.4$

Find the quotient.

11. $22.78 \div 6.7$

    3.4

12. $49.6 \div 8$

    6.2

13. $20.37 \div 3.5$

    5.82

Solve.

14. Kyle rode his bicycle a total of 48 kilometers at a rate of 8 kilometers per hour. How long did he ride?

    6 hr

15. Joanne earns $24.50 per hour as a construction worker. How much does she earn if she works 7.5 hours?

    $183.75

Name _____ LESSON 5.3

Problem Solving Strategy: Make a Table

For 1–6, use the data below. Display the data in the table at the right using intervals of 31–40, 41–50, 51–60, and 61–70.

During the basketball season, the Falcons scored the following numbers of points in their games: 63, 52, 47, 51, 60, 49, 48, 54, 61, 52, 40, 38, 57, 46, 44, 63, 70

Number of Points Scored	
31–40	//
41–50	//////
51–60	////// /
61–70	////

1. How many rows of data are in your table?

    _____ 4 rows _____

2. How many scores are greater than 40 but less than 61?

    _____ 11 scores _____

3. The Falcons won every game in which they scored more than 60 points. How many games did they play in which they scored more than 60 points?

    _____ 4 games _____

4. The team lost every game in which they did not score more than 40 points. How many games did they play in which they did not score more than 40 points?

    _____ 2 games _____

5. The Falcons' record for the season was 10 wins and 7 losses. How many games did they win in which they scored 60 points or fewer?

    _____ 6 games _____

6. With a record of 10 wins and 7 losses, how many games did the Falcons lose when they scored more than 40 points?

    _____ 5 games _____

Solve.

7. Dennis has 5 friends and wants to invite 2 of them to go to a baseball game with him and his family. How many different choices of 2 friends can Dennis make?

    _____ 10 different choices _____

8. Latifah has a project that is due on May 15. She expects the project to take her 3 weeks to complete. What is the latest date on which she could begin her project in order to be done on time?

    _____ April 24 _____

Mixed Review

Write the percent or decimal.

9. 16%   10. 5%   11. 0.55   12. 0.83   13. 0.07

   _0.16_   _0.05_   _55%_   _83%_   _7%_

PW22 Practice

Name _____

LESSON 5.4

Frequency Tables and Line Plots

Vocabulary

1. A running total of the number of people surveyed is called _____cumulative frequency_____.

2. A ____frequency table____ shows the total for each category or group in a set of data.

For 3–4, use the data in the chart at the right.

Students' Heights (cm)					
160	137	158	155	136	154
154	159	142	147	148	144
152	133	135	136	162	158
139	160	154	139	159	144
155	147	136	148	162	133

3. Find the range. ____29____

4. Make a line plot

For 5–6, use the chart at the right.

Reading Test Scores				
98	100	81	92	78
75	96	78	84	100
82	100	100	86	78

5. Find the range. ____25____

6. Make a line plot

Mixed Review

Compare the numbers. Write >, <, or = for ●.

7. 31.7 ● 37.1 ____<____

8. 72.67 ● 72.670 ____=____

9. 66.61 ● 66.16 ____>____

Solve. Use the information in the table.

10. Estimate the combined population of the four cities. ____230,000____

11. How many more people lived in Billings than in Missoula? ____39,991____

Population of the Four Largest Cities in Montana in 1999	
City	Population
Billings	91,195
Great Falls	57,758
Missoula	51,204
Butte-Silver Bow	34,051

Practice PW23

Name _____

LESSON 5.5

Measures of Central Tendency

Vocabulary

Write the correct letter from Column 2.

Column 1   **Column 2**

__b__ 1. mean   a. number that appears most often in a group of numbers

__c__ 2. median   b. sum of a group of numbers divided by the number of addends

__a__ 3. mode   c. middle number in a group of numbers arranged in order

---

Complete the table.

	Data	Mean	Median	Mode
4.	12, 15, 11, 15, 13, 10, 15	13	13	15
5.	68, 74, 71, 69, 74, 78, 70	72	71	74
6.	7.6, 6.2, 6.0, 6.2, 8.1, 6.7	6.8	6.45	6.2
7.	168, 212, 146, 195, 200, 156	179.5	181.5	(none)

For 8–10, use the table below.

Test	1	2	3	4	5	6
Score	91	84	96	89	93	84

8. Find the mean.   9. Find the median.   10. Find the mode.

       89.5   90   84

Test Scores									
98	88	82	91	83	76	98	100	84	90

11. Use the data above to make a line plot. Use your line plot to find the median and mode. Check students' plots; median: 89; mode: 98.

Mixed Review

Write the numbers in order from least to greatest.

12. 218.4, 284.1, 241.8, 214.8   13. 6.17, 6.71, 6.107, 6.701

    214.8, 218.4, 241.8, 284.1   6.107, 6.17, 6.701, 6.71

PW24 Practice

Name _____

LESSON 5.6

Outliers and Additional Data

Use the following data for 1–2.

The first 6 packages that were checked in at an airline ticket counter when it opened for business weighed 15 pounds, 21 pounds, 19 pounds, 14 pounds, 18 pounds, and 15 pounds.

1. Find the mean, median, and mode of the weights of the first 6 packages that were checked in at the counter.

   _____17 lb; 16.5 lb; 15 lb_____

2. The next package checked in weighed 66 pounds. Find the mean, median, and mode of the weights of the 7 packages.

   _____24 lb; 18 lb; 15 lb_____

3. Some friends in the school chorus compared the number of siblings they had. Two had 3 siblings, three had 2 siblings, 6 had one sibling, and 1 had no siblings. What were the mean, median, and mode number of siblings for the group of friends?

   _____1.5 siblings, 1 sibling, 1 sibling_____

4. Refer to Exercise 3. Suppose another student who has 9 siblings joins the discussion. Of the three measures of central tendency (mean, median, and mode), which measure is affected the most by including the new data value of 9 siblings?

   _____the mean_____

Use the following data for 5–6.

A survey of stores found the following prices for a popular type of backpack: $17, $19, $21, $21, $20, $18, and $24.

5. What were the mean, median, and mode of the prices for the backpack that were found during the survey?

   _____$20; $20; $21_____

6. One of the stores charging $21 reduced its price to $14 for one day. What are the new mean, median, and mode?

   _____$19; $19; no mode_____

Mixed Review

Add or subtract.

7. 6.8
    $+\,7.9$
    _____
    14.7

8. 7.03
    $+\,3.89$
    _____
    10.92

9. 15.4
    $-\,6.7$
    _____
    8.7

10. 45.04
    $-\,27.5$
    _____
    17.54

11. 12.4
    64.8
    $+\,15.7$
    _____
    92.9

12. 83.04
    $-\,59.23$
    _____
    23.81

13. 102.5
    8.7
    $+\,76.04$
    _____
    187.24

14. 234.19
    $-\,56.97$
    _____
    177.22

Estimate. Possible estimates are given.

15. 19.8×6.3

    _____120_____

16. $51.2 - 19.9$

    _____30_____

17. $8.9 + 24.2 + 16.7$

    _____50_____

Name _____

LESSON 5.7

Data and Conclusions

Write *yes* or *no* to tell whether the conclusion is valid.
Explain your answer.

1. A random sample of 100 middle school students were asked whether they think speed limits should be increased. Almost all of them believed that they should be. You conclude that drivers in general want higher speed limits.

   No. The sample is not representative of all drivers; in fact, the students are not members of the population of drivers.

2. At the music store where you buy CDs, the most popular type of music is rhythm and blues. You tell your friends that rhythm and blues must be the most popular type of music in the country.

   No. The store in which you shop may not be typical of stores around the country.

3. Your teacher tells you that in your class of 24 students, there are 2 student birthdays each month. You decide that in the entire school, student birthdays are distributed evenly throughout the year.

   Yes. Your class may or may not be a representative sample of the school population, but birthdays would probably be fairly evenly distributed for any random population.

4. You ask the first 60 students in line at the student cafeteria how they come to school. All but 15 students say they ride a school bus. You conclude that most students come to school by bus.

   Yes. The students in line probably represent a random sample of the students in the school.

Mixed Review

Find the value.

5. $5 + 5 \times 4$ ___25___
6. $6 + (4 \times 3)$ ___18___
7. $20 \div 2 + 8$ ___18___
8. $(14 - 6) \times 2$ ___16___
9. $35 \div (15 - 8)$ ___5___
10. $10 + (8 \times 3)$ ___34___

Find the mean, median, and mode.

11. 412, 387, 297, 343

    359.75; 365; no mode

12. 11, 14, 19, 14, 17, 16, 7

    14; 14; 14

13. 6.7, 6.7, 7.6, 4.2, 7.6

    6.56; 6.7; 6.7 and 7.6

PW26 Practice

Name _____

LESSON 6.1

Make and Analyze Graphs

Tell if you would use a bar, line, or circle graph to display the data.

1. The amounts of time you spend in your classes in one day.

   _____ circle or bar graph _____

2. The amounts of money you spend every day for two weeks.

   _____ line or bar graph _____

3. The number of students who play different musical instruments.

   _____ circle graph _____

4. The weights of 8 different pets.

   _____ bar graph _____

5. Make a multiple-bar graph of the homework data below.

   Hours Spent Doing Homework

Name	Science	Math
Nigel	2.5 hr	0.5 hr
Marty	1 hr	1.5 hr
Julie	0.75 hr	2 hr
Luis	1.25 hr	1 hr

6. Make a multiple-line graph of the temperature data below.

   Average Low Temperature

Year	Jan	Feb	Mar
1998	⁻5°F	5°F	8°F
1999	6°F	3°F	12°F
2000	⁻10°F	9°F	16°F
2001	12°F	⁻5°F	15°F

Check students' graphs.

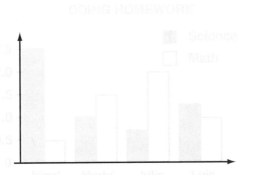

7. Gretchen researched the number of new students who came to her school during 5 months of the school years 2000 and 2001. Her data are shown at the right. What kind of graph would you use to display the data? Explain.

   Number of New Students

	Sept	Oct	Jan	Feb	Mar
2000	35	12	10	6	9
2001	5	23	14	0	12

   _____ multiple-bar graph; to compare two sets of data _____

Mixed Review

Estimate. Possible estimates are given.

8. $71.3 + 68.6 + 69.7$

   _____ 210 _____

9. $284.17 \div 7.24$

   _____ 40 _____

10. 979.88×31.05

    _____ 30,000 _____

Practice PW27

Name _____

LESSON 6.2

Find Unknown Values

Sarina kept a record of her after-school earnings.

Number of Weeks Worked	1	2	3	4
Total Saved	$16	$30	$44	$58

1. Use the data in the table to make a line graph. Use the line graph to estimate how much Sarina will have saved after working for 5 weeks.

   Check students' graphs; about $70.

2. Use logical reasoning and arithmetic to find how much Sarina will have saved after working for 5 weeks. _____$70_____

3. Use the line graph to estimate how many weeks Sarina will need to work in order to save $98. _____about 7 weeks_____

4. Use logical reasoning and arithmetic to find how many weeks Sarina will have to work to save $98. _____7 weeks_____

A train averages 60 mi per hr while traveling between New York City and Chicago.

Time (hr)	1	2	3	4
Distance (mi)	60	120	180	240

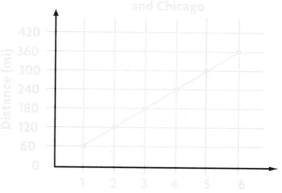

5. Use the data in the table to make a line graph. Use the line graph to estimate how long it will take the train to travel 360 mi.

   Check students' graphs; about 6 hr.

6. Use logical reasoning and arithmetic to find how long it will take the train to travel 360 mi. _____6 hr_____

7. Use the formula $d = rt$ to find how long it will take the train to travel 480 mi. _____8 hr_____

Mixed Review

Use mental math to find the value.

8. $59 + 16$ __75__   9. $63 - 21$ __42__   10. $89 - 54$ __35__

Compare the numbers. Write $<$, $>$, or $=$ for each ●.

11. 0.547 ● 0.574 __<__   12. 3.61 ● 3.16 __>__   13. 68.90 ● 68.9 __=__

PW28 Practice

Name _____

LESSON 6.3

Stem-and-Leaf Plots and Histograms

Tell whether a bar graph or a histogram is more appropriate.

1. number of fish caught at different times of day

   _____

2. average monthly phone bill for every month of one year

   _____

3. number of shoppers in a store during 3 different time intervals

   _____

Make a stem-and-leaf plot of each set of data.

4. Janet's math test scores:
   95, 83, 78, 91, 75, 85, 91, 98, 80

5. Raoul's golf scores:
   79, 85, 82, 86, 90, 94, 83, 85, 79, 91

For 6–7, use the table below.

Campers at Day Camp

Age	5–7	8–10	11–13	14–16
Number	6	11	18	9

6. Make a histogram.

7. How would the number of campers in each group change if you used 5 groups instead of 4 groups?

   _____
   _____

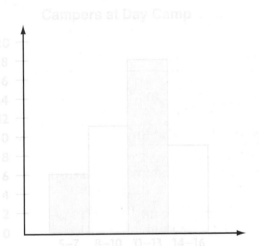

For 8–9, use the histogram at the right.

8. During which time period did the most flights arrive?

   _____

9. How many flights arrived after 11:00?

   _____

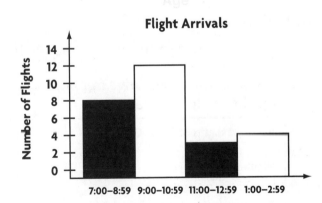

Mixed Review

10. Bill has 180 baseball cards. He has 3 times as many infielders as outfielders. How many of each does he have?

    _____

11. Tim gave a clerk $20.00 for a book and received $3.85 in change. How much did the book cost?

    _____

Practice PW29

Name _____

LESSON 6.5

Box-and-Whisker Graphs

For 1–3, use the box-and-whisker graph below.

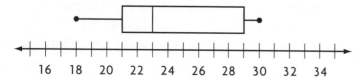

1. What is the median? ___23___

2. What are the lower and upper quartiles? ___21; 29___

3. What are the lower and upper extremes and the range? ___18; 30; 12___

For 4–8, use the data in the chart below.

Lengths of Phone Calls (in min)									
17	21	16	22	24	26	18	28	25	29
21	18	14	23	25	18	26	24	22	23

4. What is the median? ___22.5___

5. What are the lower and upper quartiles? ___18; 25___

6. What are the lower and upper extremes and the range? ___14; 29; 15___

7. Make a box-and-whisker graph.   Check students' graphs.

8. What fractional part of the data is less than 25 minutes? ___7/10___

Mixed Review

For 9–10, use the data in the chart above for 4–8.

9. Complete the cumulative frequency table below for the data.

Lengths of Phone Calls		
Minutes	Frequency	Cumulative Frequency
11–15	1	1
16–20	5	6
21–25	10	16
26–30	4	20

10. Make a line plot for the data.
    Check students' graphs.

PW30 Practice

Name _____

LESSON 6.6

Analyze Graphs

Renee asked each student in her math class the following question: "Would you rather have some great vanilla ice cream or would you prefer chocolate or strawberry?"

For 1–2, use the graph at the right, which shows the results of her survey.

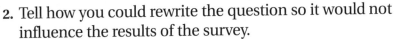

1. Could the way Renee asked the question have influenced her classmates' answers? Explain.

   Yes. The question is biased and could
   lead people to choose vanilla ice cream.

2. Tell how you could rewrite the question so it would not influence the results of the survey.

   Possible answer: "Which ice cream flavor do you
   prefer, chocolate, strawberry, or vanilla?"

A television network used the graph at the right. The network wanted to convince viewers that one of its shows, Show A, was far more popular than one of its competitors' shows, Show B, which airs at the same time.

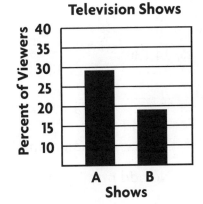

3. The bar for Show A is about how many times as high as the bar for Show B?

   about twice as high

4. Does twice the percent of the viewing audience watch Show A as watches Show B? ___no___

5. How can you change the graph so that it is not misleading?

   Adjust the scale to start at zero and have equal intervals.

Mixed Review

During one day at an airport, an airline experienced flight delays of the following numbers of minutes: 5, 7, 5, 10, 15, 15, 20, 91.

6. Find the mean length of all the flight delays. ___21 min___

7. Find the mean of the delays if the outlier is not included. ___11 min___

Evaluate each expression.

8. $g + 1.7$ for $g = 3.3$

   ___5___

9. $5y$ for $y = 1.8$

   ___9___

10. $p - 4.9$ for $p = 11$

    ___6.1___

Practice   PW31

LESSON 7.1

Name _____

Divisibility

Tell whether each number is divisible by 2, 3, 4, 5, 6, 8, 9, or 10.

1. 30
   2, 3, 5, 6, 10
2. 24
   2, 3, 4, 6, 8
3. 115
   5
4. 240
   2, 3, 4, 5, 6, 8, 10

5. 486
   2, 3, 6, 9
6. 235
   5
7. 279
   3, 9
8. 801
   3, 9

9. 145
   5
10. 650
    2, 5, 10
11. 736
    2, 4, 8
12. 1,200
    2, 3, 4, 5, 6, 8, 10

13. 207
    3, 9
14. 723
    3
15. 2,344
    2, 4, 8
16. 868
    2, 4

17. 694
    2
18. 4,464
    2, 3, 4, 5, 6, 8, 9
19. 3,894
    2, 3, 6
20. 306
    2, 3, 6, 9

21. 836
    2, 4
22. 5,962
    2
23. 2,388
    2, 3, 4, 6
24. 792
    2, 3, 4, 5, 6, 8, 9

25. 14,730
    2, 3, 5, 6, 10
26. 24,456
    2, 3, 4, 6, 8
27. 7,677
    3, 9
28. 34,248
    2, 3, 4, 6, 8

For 29–31 write *T* or *F* to tell whether each statement is true or false. If it is false, give an example that shows it is false.

29. No odd number is divisible by 2. __T__

30. All numbers that are divisible by 4 are also divisible by 2. __T__

31. All numbers that are divisible by 3 are also divisible by 6. __F; 9__

32. A number is between 40 and 50 and is divisible by both 3 and 4.

    What is the number? __48__

Mixed Review

Add or subtract mentally.

33. 451 − 71
    380
34. 898 − 196
    702
35. 109 + 46 + 54
    209

PW32 Practice

Name _____

Prime Factorization

Vocabulary

1. Write *true* or *false*. Prime factorization renames a composite number as the product of prime factors. __true__

Use division or a factor tree to find the prime factorization.

2. 28 __2 × 2 × 7__

3. 50 __2 × 5 × 5__

4. 76 __2 × 2 × 19__

5. 108 __2 × 2 × 3 × 3 × 3__

6. 55 __5 × 11__

7. 120 __2 × 2 × 2 × 3 × 5__

8. 92 __2 × 2 × 23__

Write the prime factorization in exponent form.

9. 27 __3 × 3 × 3; 3^3__

10. 100 __2 × 2 × 5 × 5; $2^2 × 5^2$__

11. 780 __2 × 2 × 3 × 5 × 13; $2^2 × 3 × 5 × 13$__

Solve for *n* to complete the prime factorization.

12. $n × 17 = 51$ __n = 3__

13. $3^n × 2 = 18$ __n = 2__

14. $2 × 2 × 2 × n = 40$ __n = 5__

Mixed Review

For 15–18, find the mean, median, and mode.

15. 28, 35, 40, 28, 33, 36, 39, 31 __33.75; 34; 28__

16. 7, 7, 8, 9, 6, 6, 7, 10, 10, 9 __7.9; 7.5; 7__

17. 428, 472, 510, 386, 440 __447.2; 440; no mode__

18. 78, 80, 95, 83, 100, 89, 88, 95 __88.5; 88.5; 95__

19. There are 24 students in Mrs. Garcia's class. She wants to divide the class evenly into groups of at least 4 students. Write the ways in which she can divide the class.

__2 groups of 12, 3 groups of 8, 4 groups of 6, or 6 groups of 4__

Practice PW33

Name _____

LESSON 7.3

Least Common Multiple and Greatest Common Factor

Vocabulary

Complete.

1. The smallest of the common multiples is called the

    _____least common multiple, or LCM_____.

2. The largest of the common factors is called the

    _____greatest common factor, or GCF_____.

List the first five multiples of each number.

3. 9
    __9, 18, 27, 36, 45__

4. 14
    __14, 28, 42, 56, 70__

5. 22
    __22, 44, 66, 88, 110__

Find the LCM of each set of numbers.

6. 12, 18   __36__
7. 7, 14   __14__
8. 16, 20   __80__
9. 4, 5, 6   __60__
10. 2, 6, 7   __42__

Find the GCF of each set of numbers.

11. 15, 45   __15__
12. 6, 14   __2__
13. 24, 40   __8__
14. 8, 12, 52   __4__
15. 16, 24, 32   __8__

Find a pair of numbers for each set of conditions.

16. The LCM is 35.
    The GCF is 7.
    __7 and 35__

17. The LCM is 36.
    The GCF is 1.
    __4 and 9__

18. The LCM is 120.
    The GCF is 10.
    __30 and 40__

Mixed Review

Determine whether each number is divisible by 2, 3, 4, 5, 6, 8, 9, or 10.

19. 72 __2, 3, 4, 6, 8, 9__

20. 80 __2, 4, 5, 8, 10__

21. 324 __2, 3, 4, 6, 9__

22. 1,500 __2, 3, 4, 5, 6, 10__

Solve for n to complete the prime factorization.

23. $2 \times n \times 7 = 42$ __n = 3__
24. $3^2 \times n = 63$ __n = 7__
25. $5 \times 7 \times n = 385$ __n = 11__

Find the quotient.

26. $24.14 \div 7.1$ __3.4__
27. $17.29 \div 3.8$ __4.55__
28. $65.024 \div 6.35$ __10.24__

PW34 Practice

Name _____

LESSON 7.4

Problem Solving Strategy: Make an Organized List

Solve the problem by making an organized list.

1. Jack and Ashley begin jogging around a quarter-mile track at the same time. Ashley takes 2 minutes to complete each lap and Jack takes 3 minutes. How many laps will each have run the first time they are side-by-side again at the point where they began?

   _____Ashley: 3 laps; Jack: 2 laps_____

2. Terrence is taking two medications for his flu. He begins taking them both at 10:00 P.M. on Tuesday. If he takes one every 8 hours and the other every 10 hours, on what day and at what time will he take the two medications together again?

   _____2:00 P.M. on Thursday_____

3. A large high school has a marching band with 64 woodwind players and 72 brass players. All members of the band line up in rows of equal size. Only musicians playing the same instruments are in each row. What is the greatest number of musicians who can be in one row?

   _____8 musicians_____

4. Brice plays in a basketball league. In his last game, he scored more than 20 but fewer than 30 points by making a combination of 2- and 3-point shots. If he made 5 more 2-point shots than 3-point shots, how many of each type did he make?

   _____8 2-point shots, 3 3-point shots_____

5. Aki is buying franks and buns for a field trip. She sees franks in packages of 6 and buns in packages of 8. There are 70 people going on the trip. What is the least number of each she can buy so there are franks and buns for everyone, with no extra packages?

   _____9 packages of buns,_____
   _____12 packages of franks_____

6. Kiona has 235 CDs. She is buying CD holders for her collection. The two types that she likes hold 20 CDs and 12 CDs each. She wants to buy the same number of each type. What is the least number of each type of CD holder that Kiona will have to buy to hold her entire CD collection?

   _____8 of each type_____

Mixed Review

Estimate the sum or difference. Possible estimates are given.

7. $80 + 31 + 87$

   _____200_____

8. $710 - 189$

   _____500_____

9. $1{,}208 + 877 + 439$

   _____2,500_____

10. $7{,}151 - 2{,}993$

    _____4,000_____

11. $67 + 123 + 804$

    _____1,000_____

12. $920 - 592$

    _____300_____

Practice   PW35

Name _____

LESSON 8.1

Equivalent Fractions and Simplest Form

Vocabulary

Complete.

1. When the numerator and denominator of a fraction have no common factor other than 1, the fraction is in _____simplest form_____.

2. Fractions that name the same amount or the same part of a whole are called _____equivalent fractions_____.

Write the factors common to the numerator and denominator.

3. $\frac{8}{32}$ 4. $\frac{10}{50}$ 5. $\frac{2}{13}$ 6. $\frac{14}{49}$ 7. $\frac{1}{19}$
   1, 2, 4, 8   1, 2, 5, 10   1   1, 7   1

8. $\frac{12}{18}$ 9. $\frac{25}{75}$ 10. $\frac{15}{40}$ 11. $\frac{9}{54}$ 12. $\frac{6}{33}$
   1, 2, 3, 6   1, 5, 25   1, 5   1, 3, 9   1, 3

Write the fraction in simplest form.

13. $\frac{9}{36}$ 14. $\frac{15}{50}$ 15. $\frac{11}{121}$ 16. $\frac{15}{36}$ 17. $\frac{14}{28}$
    $\frac{1}{4}$ $\frac{3}{10}$ $\frac{1}{11}$ $\frac{5}{12}$ $\frac{1}{2}$

18. $\frac{30}{66}$ 19. $\frac{63}{72}$ 20. $\frac{27}{81}$ 21. $\frac{25}{65}$ 22. $\frac{12}{42}$
    $\frac{5}{11}$ $\frac{7}{8}$ $\frac{1}{3}$ $\frac{5}{13}$ $\frac{2}{7}$

Complete.

23. $\frac{36}{72} = \frac{1}{\boxed{2}}$ 24. $\frac{\boxed{50}}{75} = \frac{2}{3}$ 25. $\frac{17}{\boxed{85}} = \frac{1}{5}$ 26. $\frac{63}{84} = \frac{3}{\boxed{4}}$ 27. $\frac{2}{\boxed{3}} = \frac{64}{96}$

Mixed Review

Tell whether you would use a bar, line, or circle graph to display the data.

28. The number of students in each grade at your school _____bar_____

29. A hospital patient's temperature taken each hour for 8 hours _____line_____

30. The part of each day you spend at various activities _____circle_____

PW36 Practice

Name _____

LESSON 8.2

Mixed Numbers and Fractions

Vocabulary

Complete.

1. A _____mixed number_____ has a whole-number part and a fraction part.

Write the fraction as a mixed number or a whole number.

2. $\frac{20}{5}$ 3. $\frac{19}{4}$ 4. $\frac{22}{7}$ 5. $\frac{39}{10}$ 6. $\frac{19}{10}$

7. $\frac{75}{15}$ 8. $\frac{44}{13}$ 9. $\frac{50}{7}$ 10. $\frac{63}{21}$ 11. $\frac{41}{8}$

12. $\frac{25}{6}$ 13. $\frac{72}{12}$ 14. $\frac{55}{9}$ 15. $\frac{46}{5}$ 16. $\frac{77}{11}$

Write the mixed number as a fraction.

17. $6\frac{2}{7}$ 18. $4\frac{6}{11}$ 19. $9\frac{2}{3}$ 20. $11\frac{1}{5}$ 21. $2\frac{2}{3}$

22. $7\frac{2}{9}$ 23. $12\frac{4}{5}$ 24. $4\frac{5}{8}$ 25. $8\frac{2}{3}$ 26. $13\frac{1}{2}$

Mixed Review

Write the prime factorization of each number using exponents.

27. 84 28. 72 29. 300

Write the fraction in simplest form.

30. $\frac{35}{45}$ 31. $\frac{36}{42}$ 32. $\frac{56}{72}$ 33. $\frac{22}{55}$ 34. $\frac{18}{81}$

35. $\frac{24}{30}$ 36. $\frac{16}{40}$ 37. $\frac{24}{36}$ 38. $\frac{27}{63}$ 39. $\frac{72}{88}$

Practice   PW37

Name _____   **LESSON 8.3**

Compare and Order Fractions

Compare the fractions. Write <, >, or = for each ⬤.

1. $\frac{5}{6}$ ⬤ $\frac{3}{4}$ >
2. $\frac{1}{4}$ ⬤ $\frac{1}{5}$ >
3. $\frac{2}{3}$ ⬤ $\frac{3}{8}$ >
4. $\frac{5}{8}$ ⬤ $\frac{3}{4}$ <
5. $\frac{9}{10}$ ⬤ $\frac{7}{8}$ >
6. $\frac{7}{12}$ ⬤ $\frac{3}{4}$ <
7. $\frac{13}{16}$ ⬤ $\frac{5}{6}$ <
8. $\frac{1}{7}$ ⬤ $\frac{1}{6}$ <
9. $\frac{2}{5}$ ⬤ $\frac{5}{6}$ <
10. $\frac{9}{15}$ ⬤ $\frac{3}{5}$ =
11. $\frac{4}{7}$ ⬤ $\frac{3}{5}$ <
12. $\frac{7}{8}$ ⬤ $\frac{17}{20}$ >
13. $\frac{4}{5}$ ⬤ $\frac{16}{20}$ =
14. $\frac{7}{9}$ ⬤ $\frac{2}{3}$ >
15. $\frac{1}{9}$ ⬤ $\frac{2}{3}$ <
16. $\frac{5}{9}$ ⬤ $\frac{6}{11}$ >

Use the number line to order the fractions from least to greatest.

17. $\frac{1}{6}, \frac{5}{12}, \frac{1}{3}$ _____

18. $\frac{5}{6}, \frac{7}{12}, \frac{1}{2}$ _____

19. $\frac{3}{4}, \frac{11}{12}, \frac{2}{3}$ _____

20. $\frac{2}{3}, \frac{7}{12}, \frac{5}{12}$ _____

21. $\frac{1}{2}, \frac{5}{6}, \frac{1}{6}$ _____

22. $\frac{7}{12}, \frac{1}{6}, \frac{1}{3}$ _____

Order the fractions from least to greatest.

23. $\frac{1}{4}, \frac{1}{6}, \frac{2}{5}$ _____

24. $\frac{4}{5}, \frac{2}{3}, \frac{3}{10}$ _____

25. $\frac{1}{5}, \frac{3}{8}, \frac{4}{5}$ _____

26. $\frac{7}{8}, \frac{4}{5}, \frac{9}{10}$ _____

27. $\frac{3}{4}, \frac{7}{10}, \frac{5}{7}$ _____

28. $\frac{3}{5}, \frac{1}{8}, \frac{3}{10}$ _____

Mixed Review

Find the mean, median, and mode.

29. 6, 6, 2, 4, 8, 6, 5, 3 _____

30. 23, 26, 24, 19, 31, 33 _____

31. 12, 9, 21, 11, 15, 15, 8 _____

PW38 Practice

Name _____

LESSON 8.5

Fractions, Decimals, and Percents

Write the decimal as a fraction.

1. 0.5 $\frac{5}{10}$
2. 0.14 $\frac{14}{100}$
3. 0.06 $\frac{6}{100}$
4. 0.83 $\frac{83}{100}$

5. 0.62 $\frac{62}{100}$
6. 0.317 $\frac{317}{1,000}$
7. 0.805 $\frac{805}{1,000}$
8. 0.955 $\frac{955}{1,000}$

Write as a decimal. Tell whether the decimal terminates or repeats.

9. $\frac{3}{10}$ 0.3, T
10. $\frac{6}{9}$ $0.\overline{6}$, R
11. $\frac{7}{12}$ $0.58\overline{3}$, R
12. $\frac{11}{20}$ 0.55, T

13. $\frac{7}{30}$ $0.2\overline{3}$, R
14. $\frac{9}{10}$ 0.9, T
15. $\frac{7}{15}$ $0.4\overline{6}$, R
16. $\frac{4}{11}$ $0.\overline{36}$, R

Compare. Write <, >, or = for each ●.

17. 0.24 ● $\frac{1}{4}$ <
18. 0.18 ● $\frac{7}{50}$ >
19. $\frac{4}{10}$ ● 0.44 <

20. $\frac{1}{5}$ ● 0.19 >
21. $\frac{7}{20}$ ● 0.45 <
22. $\frac{9}{20}$ ● 0.45 =

Write the fraction as a percent.

23. $\frac{3}{5}$ 60%
24. $\frac{17}{100}$ 17%
25. $\frac{4}{2}$ 200%
26. $\frac{1}{500}$ 0.2%

27. $\frac{9}{25}$ 36%
28. $\frac{7}{5}$ 140%
29. $\frac{6}{40}$ 15%
30. $\frac{17}{20}$ 85%

Mixed Review

Estimate. Possible estimates are given.

31. 56.09 ÷ 7.1 8
32. 64.1 − 13.9 50
33. 97.6 ÷ 9.8 10
34. $1.79 − $0.82 $1.00

35. 188.2 × 21.3 4,000
36. 602.5 + 102.4 700
37. $49.34 × 5 $250
38. 711.2 + 798.5 1,500

Evaluate the expression.

39. 6 + 4 × 3 18
40. 18 − 6 + 2 14
41. (10 × 3) ÷ 6 5

Practice PW39

Name _____   LESSON 9.1

Estimate Sums and Differences

Use the number line to tell whether the
fraction is closest to 0, $\frac{1}{2}$, or 1. Write *close
to 0*, *close to $\frac{1}{2}$*, or *close to 1*.

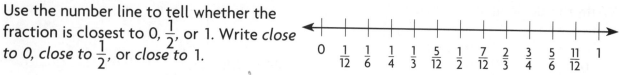

1. $\frac{2}{3}$

2. $\frac{1}{12}$

3. $\frac{11}{12}$

4. $\frac{1}{3}$

_____   _____   _____   _____

Estimate the sum or difference.

5. $\frac{4}{5} + \frac{1}{8}$

6. $\frac{5}{6} - \frac{2}{3}$

7. $\frac{1}{10} + \frac{4}{7}$

8. $\frac{3}{4} + \frac{9}{10}$

_____   _____   _____   _____

9. $5\frac{3}{5} + 1\frac{7}{8}$

10. $6\frac{10}{11} - 3\frac{1}{15}$

11. $4\frac{2}{9} + 3\frac{6}{7}$

12. $8\frac{7}{9} - 3\frac{11}{12}$

_____   _____   _____   _____

13. $\frac{1}{7} + \frac{9}{11}$

14. $\frac{4}{7} - \frac{1}{2}$

15. $\frac{5}{11} + \frac{8}{10}$

16. $\frac{3}{4} - \frac{1}{9}$

_____   _____   _____   _____

17. $9\frac{11}{14} - 1\frac{7}{10}$

18. $5\frac{2}{3} + 4\frac{1}{6}$

19. $8\frac{5}{7} - 3\frac{4}{5}$

20. $6\frac{7}{16} - 4\frac{5}{6}$

_____   _____   _____   _____

Use a range to estimate each sum or difference.

21. $7\frac{3}{4} - 1\frac{1}{12}$

22. $12\frac{1}{4} - 4\frac{1}{8}$

23. $13\frac{8}{9} + 1\frac{1}{4}$

24. $6\frac{2}{15} + 4\frac{3}{4}$

_____   _____   _____   _____

Mixed Review

Write the fraction in simplest form.

25. $\frac{15}{20}$ _____

26. $\frac{16}{28}$ _____

27. $\frac{48}{96}$ _____

28. $\frac{28}{36}$ _____

29. $\frac{5}{45}$ _____

30. $\frac{8}{32}$ _____

31. $\frac{36}{63}$ _____

32. $\frac{25}{125}$ _____

Evaluate the expression for $m = 8$ and $n = 3$.

33. $4 + m \div 2$ _____

34. $6 \times n + 7$ _____

35. $15 - n \times 2$ _____

PW40   Practice

Name _____

LESSON 9.3

Add and Subtract Fractions

Use the LCD to rewrite the problem by using equivalent fractions.

1. $\frac{3}{8} + \frac{1}{2}$ 2. $\frac{3}{4} - \frac{1}{6}$ 3. $\frac{2}{3} + \frac{4}{5}$ 4. $\frac{8}{9} - \frac{1}{3}$ 5. $\frac{1}{4} + \frac{3}{7}$

$\frac{3}{8} + \frac{4}{8}$ $\frac{9}{12} - \frac{2}{12}$ $\frac{10}{15} + \frac{12}{15}$ $\frac{8}{9} - \frac{3}{9}$ $\frac{7}{28} + \frac{12}{28}$

Write the sum or difference in simplest form. Estimate to check.

6. $\frac{1}{2} + \frac{1}{5}$ 7. $\frac{6}{7} - \frac{1}{4}$ 8. $\frac{9}{10} - \frac{3}{5}$ 9. $\frac{7}{8} - \frac{1}{2}$ 10. $\frac{3}{4} + \frac{5}{8}$

$\frac{7}{10}$ $\frac{17}{28}$ $\frac{3}{10}$ $\frac{3}{8}$ $1\frac{3}{8}$

11. $\frac{4}{5} - \frac{1}{3}$ 12. $\frac{5}{8} + \frac{1}{10}$ 13. $\frac{1}{2} - \frac{1}{6}$ 14. $\frac{7}{10} + \frac{1}{4}$ 15. $\frac{5}{6} + \frac{1}{3}$

$\frac{7}{15}$ $\frac{29}{40}$ $\frac{1}{3}$ $\frac{19}{20}$ $1\frac{1}{6}$

16. $\frac{11}{12} - \frac{1}{4}$ 17. $\frac{3}{10} + \frac{1}{2}$ 18. $\frac{3}{4} + \frac{1}{12}$ 19. $\frac{6}{7} - \frac{1}{3}$ 20. $\frac{4}{5} - \frac{1}{6}$

$\frac{2}{3}$ $\frac{4}{5}$ $\frac{5}{6}$ $\frac{11}{21}$ $\frac{19}{30}$

21. $\frac{3}{4} + \frac{1}{2}$ 22. $\frac{2}{3} - \frac{3}{8}$ 23. $\frac{3}{5} + \frac{1}{15}$ 24. $\frac{13}{14} - \frac{2}{7}$ 25. $\frac{1}{3} - \frac{1}{5}$

$1\frac{1}{4}$ $\frac{7}{24}$ $\frac{2}{3}$ $\frac{9}{14}$ $\frac{2}{15}$

26. $\frac{7}{10} - \frac{2}{5}$ 27. $\frac{1}{7} + \frac{1}{2}$ 28. $\frac{7}{12} - \frac{1}{4}$ 29. $\frac{7}{15} - \frac{2}{5}$ 30. $\frac{2}{5} + \frac{1}{3}$

$\frac{3}{10}$ $\frac{9}{14}$ $\frac{1}{3}$ $\frac{1}{15}$ $\frac{11}{15}$

31. $\frac{4}{9} + \frac{1}{2}$ 32. $\frac{2}{3} - \frac{2}{7}$ 33. $\frac{5}{8} + \frac{1}{3}$ 34. $\frac{2}{3} + \frac{1}{9}$ 35. $\frac{5}{6} - \frac{1}{2}$

$\frac{17}{18}$ $\frac{8}{21}$ $\frac{23}{24}$ $\frac{7}{9}$ $\frac{1}{3}$

Mixed Review

Find the mean, median, and mode.

36. 57, 71, 50, 57, 53, 60

     58, 57, 57

37. 21, 25, 29, 18, 31, 27, 24

     25, 25, no mode

Find the quotient.

38. $26.98 \div 3.8$ 7.1   39. $1.365 \div 0.07$ 19.5   40. $174.08 \div 27.2$ 6.4

Practice PW41

Name _____

LESSON 9.4

Add and Subtract Mixed Numbers

Draw a diagram to find each sum or difference. Write the answer in simplest form. Check students' diagrams.

1. $1\frac{2}{5} + 1\frac{2}{5}$ _____
2. $2\frac{3}{8} - 1\frac{1}{4}$ _____
3. $2\frac{1}{6} + 1\frac{1}{3}$ _____

4. $3\frac{1}{2} - 1\frac{1}{4}$ _____
5. $2\frac{3}{8} + 1\frac{1}{2}$ _____
6. $2\frac{2}{3} - 1\frac{1}{6}$ _____

Write the sum or difference in simplest form. Estimate to check.

7. $1\frac{1}{5} + 1\frac{1}{4}$ _____
8. $2\frac{1}{2} - 1\frac{1}{8}$ _____
9. $8\frac{5}{12} - 1\frac{1}{4}$ _____

10. $1\frac{1}{6} + 2\frac{2}{3}$ _____
11. $4\frac{3}{4} - 2\frac{3}{8}$ _____
12. $2\frac{1}{2} + 4\frac{4}{5}$ _____

13. $5\frac{7}{9} - 3\frac{2}{3}$ _____
14. $4\frac{3}{5} - 3\frac{1}{10}$ _____
15. $1\frac{1}{6} + 4\frac{3}{4}$ _____

16. $7\frac{1}{3} - 2\frac{1}{4}$ _____
17. $5\frac{5}{6} - 1\frac{2}{3}$ _____
18. $3\frac{2}{5} + 4\frac{1}{6}$ _____

19. $3\frac{1}{2} + 1\frac{5}{8}$ _____
20. $3\frac{7}{8} + 4\frac{1}{3}$ _____
21. $6\frac{5}{8} - 2\frac{2}{5}$ _____

Mixed Review

Write the fraction as a percent.

22. $\frac{1}{4}$ _____
23. $\frac{3}{10}$ _____
24. $\frac{2}{5}$ _____

25. $\frac{5}{100}$ _____
26. $\frac{10}{5}$ _____
27. $\frac{9}{50}$ _____

Write the numbers in order from least to greatest.

28. 0.303, 0.03, 0.33, 0.033

29. 11.10, 10.01, 11.01, 10.10

30. 2.292, 2.922, 2.929, 2.229

31. 0.545, 0.55, 0.445, 0.45

32. 6.626, 6.266, 6.226, 6.662

33. 7.070, 70.07, 7.007, 7.707

PW42 Practice

Name _____

LESSON 9.6

Subtract Mixed Numbers

Write the difference in simplest form. Estimate to check.

1. $8\frac{3}{4} - 6\frac{1}{2}$
2. $4\frac{1}{5} - 2\frac{7}{10}$
3. $7\frac{1}{4} - 2\frac{2}{3}$
4. $5\frac{2}{9} - 3\frac{2}{3}$

_____   _____   _____   _____

5. $3\frac{1}{5} - 2\frac{3}{10}$
6. $5\frac{3}{8} - 4\frac{1}{2}$
7. $6\frac{1}{3} - 2\frac{3}{4}$
8. $1\frac{7}{9} - 1\frac{2}{3}$

_____   _____   _____   _____

9. $4\frac{2}{3} - 1\frac{1}{2}$
10. $5\frac{4}{5} - 3\frac{1}{4}$
11. $3\frac{1}{3} - 1\frac{4}{9}$
12. $4\frac{5}{8} - 2\frac{1}{2}$

_____   _____   _____   _____

13. $5\frac{1}{6} - 3\frac{2}{3}$
14. $4\frac{3}{5} - 2\frac{7}{10}$
15. $4\frac{1}{8} - 2\frac{3}{4}$
16. $3\frac{1}{2} - 1\frac{4}{5}$

_____   _____   _____   _____

17. $5\frac{1}{4} - 2\frac{3}{8}$
18. $6\frac{1}{4} - 4\frac{2}{5}$
19. $9\frac{3}{8} - 4\frac{1}{3}$
20. $5\frac{1}{6} - 1\frac{5}{8}$

_____   _____   _____   _____

Evaluate each expression for $a = 3\frac{1}{3}$, $b = 2\frac{1}{4}$, $c = 5\frac{1}{6}$.

21. $c - a$
22. $c - b$
23. $a - b$

_____   _____   _____

Mixed Review

Write in exponential form.

24. $5 \times 5 \times 5 \times 5$ _____
25. $10 \times 10 \times 10$ _____

26. $k \times k \times k \times k \times k$ _____
27. $w \times w$ _____

Evaluate each expression.

28. $17.61 - s$ for $s = 12.18$
29. $75.6 \div v$ for $v = 6.3$
30. $5f$ for $f = 8.7$

_____   _____   _____

Practice PW43

Name _____ LESSON 9.7

Problem Solving Strategy: Draw a Diagram

Solve by drawing a diagram.

1. In the school art room the students use square tables. Each side of a table is $4\frac{1}{2}$ ft. If some of the tables are placed end-to-end, they form a rectangle with a perimeter of 36 ft. How many tables are used to make the rectangle?

   _____3 tables_____

2. The art room is on one side of the hallway with an office, a classroom, and the music room. The art room is between the classroom and the office. The classroom is between the music room and the art room. Which two rooms are on the ends of the hallway?

   _____office, music room_____

3. During art class, 2 students can sit at each side of a square table. The students decide to make a large rectangular table by placing 5 square tables end-to-end. How many students will be able to sit at this large table?

   _____24 students_____

4. Richard is cutting a hole in a wall to hold an air conditioner. The front of the air conditioner is a rectangle 26 in. wide and 16 in. high. The wall is 72 in. wide. If the air conditioner is centered in the wall, how wide will the wall be on either side of it?

   _____23 in._____

5. Cassandra is training for a charity walk between two towns. The towns are 12 mi apart. On her first day of training, she walks $4\frac{1}{2}$ mi. If she increases her distance by $1\frac{1}{2}$ mi every 3 days, how many days will it take until Cassandra has walked at least 10 mi?

   _____13 days_____

6. Marla wants to wrap a present that is in the shape of a cube. She wants to put one piece of ribbon around the top, bottom and two sides. She wants to put a second piece around the top, bottom, and other two sides. The box is $8\frac{1}{2}$ in. on each edge. What is the shortest length of ribbon she needs?

   _____68 in._____

Mixed Review

Write the number in standard form.

7. six hundred and three tenths ___600.3___

8. ninety-one hundredths ___0.91___

9. ninety and seven hundredths ___90.07___

10. eighty and nine tenths ___80.9___

Find the GCF for each set of numbers.

11. 10, 15   12. 16, 40   13. 18, 45   14. 20, 28   15. 24, 56

    ___5___   ___8___   ___9___   ___4___   ___8___

PW44 Practice

Name _____

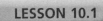

Estimate Products and Quotients

Estimate each product or quotient. Possible answers are given.

1. $4\frac{1}{4} \times 3\frac{3}{4}$ ___16___
2. $20\frac{5}{6} \div 6\frac{3}{4}$ ___3___
3. $\frac{3}{4} \times \frac{5}{6}$ ___1___

4. $\frac{3}{4} \div \frac{2}{3}$ ___1___
5. $45\frac{1}{3} \div 8\frac{2}{3}$ ___5___
6. $17\frac{2}{7} \times 1\frac{2}{7}$ ___17___

7. $2\frac{3}{5} \div \frac{2}{5}$ ___5___
8. $19 \times 6\frac{1}{3}$ ___114___
9. $2\frac{3}{4} \times 2\frac{4}{5}$ ___9___

10. $36\frac{3}{7} \div 11\frac{3}{4}$ ___3___
11. $\frac{7}{9} \times 13\frac{1}{9}$ ___13___
12. $\frac{1}{5} \div 20$ ___0___

13. $3\frac{3}{4} \div 4\frac{1}{2}$ ___1___
14. $42\frac{1}{6} \times 14\frac{4}{9}$ ___630___
15. $\frac{1}{10} \times \frac{1}{10}$ ___0___

16. $8\frac{1}{3} \times 6\frac{4}{5}$ ___56___
17. $12\frac{1}{6} \div 3\frac{2}{3}$ ___3___
18. $40\frac{2}{9} \div 7\frac{4}{5}$ ___5___

19. $10\frac{5}{6} \times 3\frac{7}{8}$ ___44___
20. $18\frac{3}{10} \div 1\frac{6}{7}$ ___9___
21. $9\frac{3}{4} \times 17\frac{1}{5}$ ___170___

Estimate to compare. Write < or > for each ●.

22. $3\frac{1}{8} \times 5$ ● $12 \div \frac{9}{10}$ ___>___
23. $6\frac{1}{2} \div 12$ ● $\frac{5}{8} \div 1\frac{2}{3}$ ___>___

24. $5\frac{2}{7} \div 1\frac{3}{8}$ ● $2\frac{1}{8} \div 3\frac{7}{8}$ ___>___
25. $3\frac{3}{4} \times 1\frac{1}{4}$ ● $31\frac{3}{4} \div 8\frac{1}{4}$ ___>___

26. $15\frac{1}{5} \div 4\frac{2}{3}$ ● $1\frac{3}{4} \div 3\frac{4}{5}$ ___>___
27. $7\frac{2}{9} \times 1\frac{5}{7}$ ● $36\frac{1}{2} \div 2\frac{7}{8}$ ___<___

Mixed Review

Write the fraction as a percent.

28. $\frac{3}{4}$
___75%___

29. $\frac{7}{10}$
___70%___

30. $\frac{1}{20}$
___5%___

31. $\frac{3}{25}$
___12%___

32. $\frac{29}{50}$
___58%___

33. $\frac{13}{10}$
___130%___

34. $\frac{1}{8}$
___12.5%___

35. $\frac{5}{8}$
___62.5%___

Practice PW45

Multiply Fractions

Make a model to find the product. Check students' models.

1. $\frac{1}{2} \times 6$ = 3
2. $\frac{2}{5} \times \frac{1}{2}$ = $\frac{1}{5}$
3. $\frac{1}{8} \times \frac{1}{2}$ = $\frac{1}{16}$
4. $10 \times \frac{1}{2}$ = 5
5. $\frac{1}{2} \times \frac{1}{3}$ = $\frac{1}{6}$

Multiply. Write the answer in simplest form.

6. $\frac{1}{4} \times \frac{1}{6}$ = $\frac{1}{24}$
7. $\frac{1}{5} \times \frac{1}{2}$ = $\frac{1}{10}$
8. $\frac{3}{8} \times \frac{1}{4}$ = $\frac{3}{32}$
9. $\frac{3}{5} \times \frac{1}{4}$ = $\frac{3}{20}$
10. $\frac{4}{5} \times \frac{1}{2}$ = $\frac{2}{5}$

11. $\frac{1}{4} \times \frac{8}{9}$ = $\frac{2}{9}$
12. $\frac{3}{4} \times \frac{2}{7}$ = $\frac{3}{14}$
13. $\frac{5}{9} \times \frac{9}{10}$ = $\frac{1}{2}$
14. $\frac{5}{6} \times \frac{2}{5}$ = $\frac{1}{3}$
15. $\frac{6}{7} \times \frac{2}{3}$ = $\frac{4}{7}$

16. $\frac{3}{4} \times \frac{8}{9}$ = $\frac{2}{3}$
17. $\frac{3}{4} \times \frac{8}{15}$ = $\frac{2}{5}$
18. $\frac{1}{6} \times \frac{8}{9}$ = $\frac{4}{27}$
19. $\frac{7}{8} \times 24$ = 21
20. $\frac{3}{8} \times \frac{1}{3}$ = $\frac{1}{8}$

21. $\frac{5}{6} \times \frac{3}{10}$ = $\frac{1}{4}$
22. $\frac{9}{10} \times \frac{2}{3}$ = $\frac{3}{5}$
23. $30 \times \frac{4}{5}$ = 24
24. $\frac{1}{2} \times \frac{12}{13}$ = $\frac{6}{13}$
25. $\frac{9}{11} \times \frac{22}{27}$ = $\frac{2}{3}$

Compare. Write <, >, or = for ●.

26. $\frac{1}{2} \times \frac{2}{3}$ ● $\frac{2}{3}$ $<$
27. $\frac{3}{4} \times 8$ ● 6 $=$
28. $\frac{1}{4} \times 4$ ● $\frac{1}{4}$ $>$

Mixed Review

Write each mixed number as a fraction.

29. $4\frac{2}{5}$ = $\frac{22}{5}$
30. $6\frac{3}{7}$ = $\frac{45}{7}$
31. $2\frac{8}{11}$ = $\frac{30}{11}$
32. $5\frac{3}{5}$ = $\frac{28}{5}$

Write each fraction as a mixed number.

33. $\frac{12}{7}$ = $1\frac{5}{7}$
34. $\frac{41}{12}$ = $3\frac{5}{12}$
35. $\frac{25}{6}$ = $4\frac{1}{6}$
36. $\frac{50}{9}$ = $5\frac{5}{9}$

Multiply Mixed Numbers

Multiply. Write your answer in simplest form.

1. $2\frac{1}{2} \times 1\frac{1}{3}$
   $3\frac{1}{3}$

2. $3\frac{1}{5} \times 2\frac{1}{2}$
   8

3. $8\frac{3}{4} \times \frac{2}{5}$
   $3\frac{1}{2}$

4. $3\frac{1}{3} \times 1\frac{1}{5}$
   4

5. $3\frac{1}{3} \times 2\frac{2}{5}$
   8

6. $1\frac{3}{4} \times \frac{3}{14}$
   $\frac{3}{8}$

7. $4\frac{2}{5} \times \frac{10}{11}$
   4

8. $\frac{6}{7} \times 2\frac{1}{10}$
   $1\frac{4}{5}$

9. $3\frac{1}{2} \times 1\frac{1}{4}$
   $4\frac{3}{8}$

10. $2\frac{3}{5} \times 1\frac{2}{3}$
    $4\frac{1}{3}$

11. $4\frac{3}{8} \times \frac{1}{2}$
    $2\frac{3}{16}$

12. $6\frac{4}{5} \times \frac{5}{8}$
    $4\frac{1}{4}$

13. $2\frac{1}{4} \times 3\frac{1}{5}$
    $7\frac{1}{5}$

14. $9\frac{1}{3} \times 1\frac{2}{7}$
    12

15. $\frac{3}{5} \times 1\frac{2}{3}$
    1

16. $12\frac{1}{3} \times 1\frac{1}{2}$
    $18\frac{1}{2}$

17. $1\frac{1}{8} \times \frac{1}{3}$
    $\frac{3}{8}$

18. $3\frac{3}{4} \times 1\frac{5}{6}$
    $6\frac{7}{8}$

19. $2\frac{2}{5} \times 1\frac{5}{8}$
    $3\frac{9}{10}$

20. $5\frac{3}{5} \times 1\frac{2}{7}$
    $7\frac{1}{5}$

Use the Distributive Property to multiply.

21. $7 \times 4\frac{1}{6}$
    $29\frac{1}{6}$

22. $1\frac{1}{4} \times 8$
    10

23. $5\frac{3}{8} \times 3$
    $16\frac{1}{8}$

24. $6 \times 2\frac{4}{5}$
    $16\frac{4}{5}$

Compare. Write <, >, or = for ●.

25. $2\frac{1}{2} \times 2\frac{3}{4}$ ● $3\frac{1}{2} \times 4$ $<$

26. $6\frac{2}{3} \times 3\frac{3}{5}$ ● $3\frac{3}{4} \times 6\frac{2}{5}$ $=$

Mixed Review

Use the data in the chart for 27–28.

Quiz Scores								
30	27	21	27	25	30	29	19	15
26	27	28	22	25	23	26	18	17

27. Make a stem-and-leaf plot of the data.

28. Use the stem-and-leaf plot to find the median and mode.
    25.5; 27

Stem	Leaves
1	5 7 8 9
2	1 2 3 5 5 6 6 7 7 7 8 9
3	0 0

Name _____

LESSON 10.5

Divide Fractions and Mixed Numbers

Write the reciprocal of the number.

1. $\frac{6}{7}$ 2. $\frac{1}{9}$ 3. 5   4. $\frac{8}{5}$ 5. $3\frac{1}{3}$

   $\frac{7}{6}$ 9 $\frac{1}{5}$ $\frac{5}{8}$ $\frac{3}{10}$

Find the quotient. Write the answer in simplest form.

6. $\frac{4}{5} \div \frac{8}{15}$ 7. $\frac{7}{10} \div \frac{1}{2}$ 8. $\frac{5}{6} \div \frac{1}{2}$ 9. $24 \div \frac{1}{2}$

   $1\frac{1}{2}$ $1\frac{2}{5}$ $1\frac{2}{3}$ 48

10. $9 \div \frac{1}{6}$ 11. $\frac{7}{9} \div \frac{2}{3}$ 12. $\frac{9}{10} \div \frac{2}{5}$ 13. $\frac{9}{20} \div \frac{3}{4}$

   54           $1\frac{1}{6}$ $2\frac{1}{4}$ $\frac{3}{5}$

14. $\frac{5}{8} \div \frac{5}{16}$ 15. $\frac{5}{6} \div \frac{2}{3}$ 16. $\frac{12}{21} \div \frac{4}{7}$ 17. $\frac{5}{8} \div \frac{1}{4}$

   2           $1\frac{1}{4}$ 1 $2\frac{1}{2}$

18. $\frac{3}{4} \div \frac{2}{3}$ 19. $\frac{5}{9} \div \frac{5}{6}$ 20. $\frac{7}{8} \div 12$ 21. $15 \div \frac{5}{9}$

   $1\frac{1}{8}$ $\frac{2}{3}$ $\frac{7}{96}$ 27

22. $\frac{5}{12} \div \frac{3}{4}$ 23. $\frac{3}{8} \div 18$ 24. $\frac{7}{10} \div 14$ 25. $24 \div \frac{4}{5}$

   $\frac{5}{9}$ $\frac{1}{48}$ $\frac{1}{20}$ 30

Use mental math to find each quotient.

26. $10 \div \frac{1}{4}$ 27. $12 \div \frac{1}{6}$ 28. $3 \div \frac{1}{10}$ 29. $15 \div \frac{1}{2}$

   40   72   30   30

Mixed Review

Find the mean, median, and mode.

30. 8, 10, 12, 11, 8, 9, 10, 10   31. 228, 209, 195, 187, 251

   9.75; 10; 10   214; 209; no mode

Compare. Write <, >, or = for ●.

32. $\frac{4}{5}$ ● $\frac{8}{9}$ 33. $\frac{5}{13}$ ● $\frac{4}{13}$ 34. $\frac{6}{15}$ ● $\frac{2}{5}$ 35. $\frac{6}{7}$ ● $\frac{14}{15}$

   <   >   =   <

PW48 Practice

Name _____

LESSON 10.6

Problem Solving Skill: Choose the Operation

Solve. Name the operations used.

1. Marie practiced piano a total of $17\frac{1}{2}$ hr last week. If she practiced the same amount of time each day, how long did she practice daily?

   $2\frac{1}{2}$ hr, division

2. Sylvan withdrew $\frac{2}{5}$ of the amount in his savings account, and spent $\frac{7}{10}$ of that money. What fraction of his total savings does he still have?

   $\frac{18}{25}$, multiplication and subtraction

3. Ike practices guitar $2\frac{1}{2}$ hr per day, but Jenn only practices $\frac{3}{4}$ hr. How much longer does Ike practice?

   $1\frac{3}{4}$ hr, subtraction

4. A painter is going to paint a wall that measures $2\frac{2}{3}$ yd by $4\frac{1}{2}$ yd. What is the area of the wall?

   12 yd², multiplication

5. José gives each of his 15 patio plants $\frac{3}{4}$ qt of water daily in warm weather. How much water does José use on his plants on a warm day?

   $11\frac{1}{4}$ qt, multiplication

6. José waters each of his 15 patio plants with $\frac{1}{2}$ qt water daily in cool weather. How much water can José expect to use on his patio plants during a cool week?

   52.5 qt, multiplication

7. Marisol rode her scooter $1\frac{1}{2}$ mi to Athena's home, then $\frac{3}{4}$ mi to Ariel's home, then $1\frac{1}{4}$ mi back to her home. How far did Marisol ride?

   $3\frac{1}{2}$ mi, addition

8. Bill can polish a car in $2\frac{3}{4}$ hr. Lara and Danny can do the same job working together in $1\frac{1}{2}$ hr. How much faster than Bill can Lara and Danny do the job when working together?

   $1\frac{1}{4}$ hr faster, subtraction

Mixed Review

Write each fraction in simplest form.

9. $\frac{5}{10}$ 10. $\frac{20}{50}$ 11. $\frac{15}{25}$ 12. $\frac{22}{32}$ 13. $\frac{21}{24}$

   $\frac{1}{2}$ $\frac{2}{5}$ $\frac{3}{5}$ $\frac{11}{16}$ $\frac{7}{8}$

Practice   PW49

Name _____

LESSON 10.7

Algebra: Fraction Expressions and Equations

Evaluate the expression.

1. $2\frac{1}{4} + x$ for $x = 2\frac{1}{8}$ _____
2. $2\frac{1}{4} + x$ for $x = \frac{1}{2}$ _____
3. $2\frac{1}{4} + x$ for $x = \frac{3}{8}$ _____

4. $y - 2\frac{3}{5}$ for $y = 5\frac{4}{5}$ _____
5. $y - 2\frac{3}{5}$ for $y = 4\frac{7}{10}$ _____
6. $y - 2\frac{3}{5}$ for $y = 6$ _____

7. $\frac{3}{5}s$ for $s = 2$ _____
8. $\frac{3}{5}s$ for $s = \frac{1}{3}$ _____
9. $\frac{3}{5}s$ for $s = 1\frac{2}{3}$ _____

10. $6\frac{2}{7}p$ for $p = \frac{1}{2}$ _____
11. $6\frac{2}{7}p$ for $p = \frac{7}{3}$ _____
12. $6\frac{2}{7}p$ for $p = 2\frac{5}{8}$ _____

13. $x \div 1\frac{1}{2}$ for $x = 4\frac{1}{2}$ _____
14. $x \div 3\frac{1}{4}$ for $x = \frac{1}{4}$ _____
15. $x \div 2\frac{1}{3}$ for $x = 2\frac{1}{3}$ _____

Use mental math to solve the equation.

16. $x + 5\frac{3}{4} = 5\frac{7}{8}$
17. $\frac{1}{2}y = \frac{1}{12}$
18. $z - 8\frac{1}{3} = 12\frac{1}{2}$
19. $w \div \frac{9}{20} = \frac{5}{9}$

_____ _____ _____ _____

20. $\frac{4}{5}n = 3$
21. $c + 4\frac{1}{3} = 7\frac{5}{6}$
22. $m - 6\frac{1}{2} = 5\frac{7}{8}$
23. $\frac{3}{4}d = 9\frac{3}{4}$

_____ _____ _____ _____

Mixed Review

Add or subtract. Write the answer in simplest form.

24. $\frac{7}{8} - \frac{3}{4}$
25. $\frac{1}{2} - \frac{1}{12}$
26. $\frac{1}{3} + \frac{1}{2}$
27. $\frac{2}{5} + \frac{9}{20}$

_____ _____ _____ _____

28. $\frac{4}{5} - \frac{3}{7}$
29. $\frac{4}{9} + \frac{3}{10}$
30. $\frac{7}{10} - \frac{1}{6}$
31. $\frac{5}{12} + \frac{8}{15}$

_____ _____ _____ _____

32. $4\frac{1}{2} + 2\frac{1}{4} + 1\frac{1}{8}$ _____
33. $4\frac{3}{8} - 2\frac{3}{4}$ _____

PW50 Practice

Name _____

LESSON 11.1

Understand Integers

Vocabulary

Complete.

1. ___Integers___ include all whole numbers and their opposites.

2. The ___absolute value___ of an integer is its distance from 0.

Write an integer to represent each situation.

3. earning 7 dollars

    ___+7___

4. digging a hole 2 feet deep

    ___−2___

5. taking 10 steps backward

    ___−10___

6. climbing up a mountain 20 feet

    ___+20___

Find the absolute value.

7. $|{-3}|$ 8. $|{+3}|$ 9. $|{-2}|$ 10. $|{-6}|$ 11. $|{+9}|$ 12. $|{-15}|$

   ___3___   ___3___   ___2___   ___6___   ___9___   ___15___

13. $|{-32}|$ 14. $|{+32}|$ 15. $|{-47}|$ 16. $|{+78}|$ 17. $|{-180}|$ 18. $|{+574}|$

   ___32___   ___32___   ___47___   ___78___   ___180___   ___574___

Write the opposite integer.

19. $^-5$ 20. $^+13$ 21. $^+21$ 22. $^-19$ 23. $^-25$ 24. $^+37$

   ___+5___   ___−13___   ___−21___   ___+19___   ___+25___   ___−37___

Mixed Review

Multiply. Write the answer in simplest form.

25. $\frac{1}{5} \times \frac{6}{7}$

    ___$\frac{6}{35}$___

26. $\frac{4}{9} \times \frac{3}{5}$

    ___$\frac{4}{15}$___

27. $\frac{4}{5} \times 30$

    ___24___

28. $2\frac{7}{10} \times \frac{2}{3}$

    ___$\frac{9}{5}$ or $1\frac{4}{5}$___

29. $3\frac{3}{4} \times 2\frac{2}{5}$

    ___9___

30. $1\frac{1}{2} \times 3\frac{1}{3}$

    ___5___

Practice   PW51

Name _____

LESSON 11.2

Rational Numbers

Use the number line to find a rational number between the two given numbers. Possible answers are given.

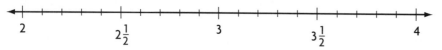

1. 2 and $2\frac{1}{2}$
   $2\frac{1}{4}$

2. $2\frac{1}{2}$ and 3
   $2\frac{3}{4}$

3. 3 and $3\frac{1}{2}$
   $3\frac{1}{4}$

4. $3\frac{1}{2}$ and 4
   $3\frac{3}{4}$

Find a rational number between the two given numbers. Possible answers are given.

5. $\frac{3}{8}$ and $\frac{4}{6}$
   $\frac{11}{24}$

6. $\frac{3}{8}$ and $\frac{2}{3}$
   $\frac{1}{2}$

7. $1\frac{7}{8}$ and $1\frac{3}{4}$
   $1\frac{13}{16}$

8. $^-3$ and $^-3\frac{1}{2}$
   $^-3\frac{1}{4}$

9. 3.1 and 3.2
   3.15

10. $^-1.7$ and $^-1.8$
    $^-1.72$

11. $^-5.6$ and $^-5.7$
    $^-5.68$

12. 3.04 and 3.05
    3.041

Write each rational number in the form $\frac{a}{b}$. Possible answers are given.

13. $3\frac{1}{2}$
    $\frac{7}{2}$

14. 0.3
    $\frac{3}{10}$

15. 0.45
    $\frac{45}{100}$, or $\frac{9}{20}$

16. 11.2
    $\frac{112}{10}$, or $\frac{56}{5}$

17. $2\frac{1}{4}$
    $\frac{9}{4}$

18. 3.15
    $\frac{315}{100}$, or $\frac{63}{20}$

19. 15
    $\frac{15}{1}$

20. 27
    $\frac{27}{1}$

21. $3\frac{1}{5}$
    $\frac{16}{5}$

22. 0.59
    $\frac{59}{100}$

23. 370
    $\frac{370}{1}$

24. $4\frac{1}{7}$
    $\frac{29}{7}$

Use the Venn diagram at the right to determine in which set or sets the number belongs.

25. 1.8
    R

26. $5\frac{2}{3}$
    R

27. 48
    all

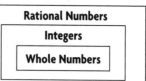

Mixed Review

Write the reciprocal of the number.

28. $\frac{6}{7}$ $\frac{7}{6}$

29. $1\frac{4}{7}$ $\frac{7}{11}$

30. 12 $\frac{1}{12}$

31. $1\frac{1}{7}$ $\frac{7}{8}$

Find the quotient. Write the answer in simplest form.

32. $\frac{2}{5} \div \frac{1}{3}$ $1\frac{1}{5}$

33. $6 \div \frac{8}{9}$ $6\frac{3}{4}$

34. $3\frac{3}{8} \div 1\frac{4}{5}$ $1\frac{7}{8}$

PW52 Practice

Name _____

LESSON 11.3

Compare and Order Rational Numbers

Compare. Write < or > for ◯.

1. 0.25 ◯ 0.4   2. $\frac{3}{8}$ ◯ 0.2   3. $^-2\frac{1}{5}$ ◯ $^-2.3$ 4. $\frac{-5}{8}$ ◯ $\frac{-3}{10}$

   <u> < </u>   <u> > </u>   <u> > </u>   <u> < </u>

5. 5 ◯ $^-2$ 6. $\frac{-7}{10}$ ◯ $\frac{4}{5}$ 7. $^-2.6$ ◯ $^-2.62$ 8. $\frac{3}{4}$ ◯ $\frac{5}{6}$

   <u> > </u>   <u> < </u>   <u> > </u>   <u> < </u>

9. $3.8 + 2.2$ ◯ $2\frac{1}{6} + 3\frac{4}{5}$ 10. $3\frac{1}{2} \times 2$ ◯ $4\frac{1}{3} + 2.8$ 11. $7\frac{1}{4} + 3\frac{1}{3}$ ◯ $1\frac{5}{6} \times 6$

   <u> > </u>   <u> < </u>   <u> < </u>

Order the rational numbers from least to greatest.

12. 2.9, $^-1.7$, $\frac{9}{3}$, $\frac{3}{4}$ 13. $\frac{-1}{5}$, $\frac{1}{9}$, $\frac{1}{10}$, $^-0.1$ 14. 0, 0.8, $^-1.4$, $^-0.6$, $\frac{3}{5}$

   <u>$^-1.7$; $\frac{3}{4}$; 2.9; $\frac{9}{3}$</u>   <u>$\frac{-1}{5}$; $^-0.1$; $\frac{1}{10}$; $\frac{1}{9}$</u>   <u>$^-1.4$; $^-0.6$; 0; $\frac{3}{5}$; 0.8</u>

15. 8.7, $^-9.2$, $^-7.3$, 6.2, $6\frac{1}{2}$, $8\frac{7}{8}$ 16. $4\frac{1}{4}$, $4\frac{3}{5}$, 4.9, 4.08, 0.49

   <u>$^-9.2$; $^-7.3$; 6.2; $6\frac{1}{2}$; 8.7; $8\frac{7}{8}$</u>   <u>0.49; 4.08; $4\frac{1}{4}$; $4\frac{3}{5}$; 4.9</u>

Order the rational numbers from greatest to least.

17. 7.3, 6, $\frac{7}{8}$, 2   18. 2.4, $^-1.4$, $^-3$, 4.7, 3.8   19. $\frac{2}{5}$, $\frac{1}{10}$, 0.5, $^-0.6$, 0.42

   <u>7.3; 6; 2; $\frac{7}{8}$</u>   <u>4.7; 3.8; 2.4; $^-1.4$; $^-3$</u>   <u>0.5; 0.42; $\frac{2}{5}$; $\frac{1}{10}$; $^-0.6$</u>

Mixed Review

Find the LCM of each set of numbers.

20. 4, 10   21. 7, 12   22. 8, 18, 24   23. 5, 15, 20

    <u> 20 </u>   <u> 84 </u>   <u> 72 </u>   <u> 60 </u>

Find the GCF of each set of numbers.

24. 12, 20   25. 16, 42   26. 15, 50, 75   27. 36, 54, 72

    <u> 4 </u>   <u> 2 </u>   <u> 5 </u>   <u> 18 </u>

Find a pair of numbers for each set of conditions. Possible answers are given.

28. The LCM is 30. The GCF is 2.   29. The LCM is 36. The GCF is 6.

    <u> 6 and 10 </u>   <u> 12 and 18 </u>

Practice   PW53

Name _____

LESSON 11.4

Problem Solving Strategy: Use Logical Reasoning

Solve the problems by using logical reasoning.

1. Tamara, Alex, Elena, and Fred entered their dogs in the county dog show. The dogs were a terrier, a setter, a golden retriever, and a Great Dane. Neither girl owned the Great Dane. Neither boy entered a setter. Tamara owns a golden retriever. What breed of dog did Elena enter in the show?

   _____setter_____

2. Bobby, Ken, Sam, and Ayesha each participate in one sport at school. They play softball, football, basketball, and soccer. Ayesha plays first base. Ken does not play football. If Sam plays soccer, what sport does Bobby participate in?

   _____football_____

3. Adel, James, Erica, and An were comparing how far they live from school. An lives only $\frac{1}{3}$ as far as Adel. James lives twice as far as Erica and 4 times as far as An. If Adel lives 9 blocks from school, how far away does Erica live?

   _____Erica: 6 blocks_____

4. Ahmed looked over his math homework problems. He saw that $\frac{1}{2}$ of the problems were about fractions, $\frac{1}{3}$ were about decimals, and the rest were about geometry. If there were 4 geometry problems, how many problems did he have in all?

   _____24 homework problems_____

5. Robert, Stanley, and Keith are brothers. Robert is 4 years younger than Stanley. Keith is 3 years older than Robert. Robert is 9 years older than his cousin Richard. If Richard is 11, how old is each brother?

   _____Robert: 20; Keith: 23; Stanley 24_____

6. Adam, Carin, Dana, and Juanita are lined up for a photograph. As the photographer looks at them, Juanita is to the right of Carin. Adam is on one end. Dana is between Carin and Adam. Give their order from left to right.

   _____Adam, Dana, Carin, Juanita_____

Mixed Review

Determine whether each number is divisible by 2, 3, 4, 5, 6, 8, 9, or 10.

7. 125   8. 336   9. 1,010   10. 249   11. 9,072

   ___5___   ___2, 3, 4, 6, 8___   ___2, 5, 10___   ___3___   ___2, 3, 4, 6, 8, 9___

Multiply. Write the answer in simplest form.

12. $\frac{1}{2} \times \frac{2}{5}$ 13. $\frac{3}{5} \times \frac{1}{3}$ 14. $\frac{5}{6} \times \frac{1}{4}$ 15. $\frac{3}{4} \times \frac{5}{6}$

   $\frac{1}{5}$ $\frac{1}{5}$ $\frac{5}{24}$ $\frac{5}{8}$

PW54 Practice

Name _____

LESSON 12.2

Add Integers

Write the addition problem modeled on the number line.

1.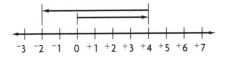

   $^+4 + {}^-6 = {}^-2$

2.

   $^-5 + {}^+10 = {}^+5$

3.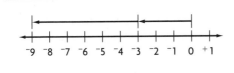

   $^-3 + {}^-6 = {}^-9$

4.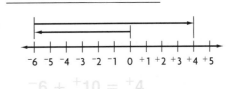

   $^-6 + {}^+10 = {}^+4$

Find the sum.

5. $^-8 + {}^-5$
   $^-13$

6. $^+14 + {}^-9$
   $^+5$

7. $^-20 + {}^-4$
   $^-24$

8. $^+31 + {}^-12$
   $^+19$

9. $^-14 + {}^-16$
   $^-30$

10. $^+35 + {}^+17$
    $^+52$

11. $^-23 + {}^-9$
    $^-32$

12. $^+39 + {}^-15$
    $^+24$

13. $^-59 + {}^-22$
    $^-81$

14. $^+47 + {}^-33$
    $^+14$

15. $^-37 + {}^-26$
    $^-63$

16. $^+49 + {}^-20$
    $^+29$

17. $^-19 + {}^-42$
    $^-61$

18. $^+17 + {}^-12$
    $^+5$

19. $^+44 + {}^-17$
    $^+27$

20. $^-64 + {}^-38$
    $^-102$

21. $^-23 + {}^+50$
    $^+27$

22. $^-31 + {}^-43$
    $^-74$

23. $^+85 + {}^-15$
    $^+70$

24. $^-59 + {}^-21$
    $^-80$

Mixed Review

Write the opposite of each number.

25. $^-12$ ___ $^+12$
26. $^+81$ ___ $^-81$
27. $^-54$ ___ $^+54$
28. $^-17$ ___ $^+17$

Find the absolute value.

29. $|{-45}|$ ___ 45
30. $|{^+101}|$ ___ 101
31. $|{^+310}|$ ___ 310
32. $|{-287}|$ ___ 287

Write each rational number in the form $\frac{a}{b}$. Possible answers are given.

33. $6\frac{7}{10}$ ___ $\frac{67}{10}$
34. $^-9\frac{1}{8}$ ___ $\frac{-73}{8}$
35. $^-1.59$ ___ $\frac{-159}{100}$
36. 4.03 ___ $\frac{403}{100}$

Practice   PW55

Name _____

LESSON 12.4

Subtract Integers

Use the number line to find the difference.

1. $^-6 - {^-9} = {^-6} + {^+9} =$ __+3__

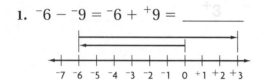

2. $^-4 - {^+5} = {^-4} + {^-5} =$ __−9__

3. $^-6 - {^+5} = {^-6} + {^-5} =$ __−11__

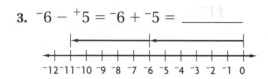

4. $^-3 - {^+7} = {^-3} + {^-7} =$ __−10__

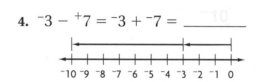

Find the difference.

5. $^+8 - {^-9}$	6. $^-14 - {^-6}$	7. $^+12 - {^-9}$	8. $^+6 - {^-2}$
+17	−8	+21	+8

9. $^+10 - {^-3}$	10. $^+11 - {^-9}$	11. $^-14 - {^-7}$	12. $^-9 - {^+3}$
+13	+20	−7	−12

13. $^-11 - {^-9}$	14. $^-9 - {^+4}$	15. $^-13 - {^+5}$	16. $^-13 - {^+2}$
−2	−13	−18	−15

17. $^-19 - {^+7}$	18. $^+16 - {^+12}$	19. $^+17 - {^-11}$	20. $^-18 - {^-9}$
−26	+4	+28	−9

21. $^+15 - {^-14}$	22. $^-19 - {^+13}$	23. $^-21 - {^+6}$	24. $^-20 - {^-8}$
+29	−32	−27	−12

Mixed Review

Find a rational number between the two given numbers. Possible answers are given.

25. 8.3 and 8.26	26. $^-4\frac{1}{2}$ and $^-4\frac{1}{3}$	27. $^-\frac{3}{8}$ and $^-0.4$	28. $^-1.9$ and $^-1\frac{3}{4}$
8.28	$^-4\frac{5}{12}$	$^-0.38$	$^-1\frac{7}{8}$

Compare. Write < or > for each ●.

29. $\frac{2}{3}$ ● $\frac{4}{5}$	30. $^-1.4$ ● $^-1\frac{3}{8}$	31. $\frac{3}{4}$ ● 0.7	32. $^-5.5$ ● $^-5.6$
<	<	>	>

PW56 Practice

Name _____

LESSON 12.5

Multiply and Divide Integers

Find the product or quotient.

1. ⁻3 × 7
   ⁻21

2. 8 × ⁻3
   ⁻24

3. ⁻14 ÷ ⁻2
   7

4. 24 ÷ ⁻3
   ⁻8

5. ⁻150 ÷ 25
   ⁻6

6. 36 ÷ 9
   4

7. ⁻80 ÷ ⁻4
   20

8. 75 ÷ ⁻25
   ⁻3

9. ⁻130 ÷ ⁻5
   26

10. ⁻4 × 6
    ⁻24

11. 9 × ⁻6
    ⁻54

12. ⁻6 × ⁻7
    42

13. ⁻12 × ⁻2
    24

14. 90 ÷ ⁻5
    ⁻18

15. 160 ÷ 16
    10

16. ⁻88 ÷ 11
    ⁻8

17. 42 ÷ 3
    14

18. ⁻70 ÷ ⁻7
    10

19. ⁻4 × 25
    ⁻100

20. ⁻35 × ⁻2
    70

21. ⁻5 × 12
    ⁻60

22. ⁻14 × ⁻7
    98

23. ⁻200 ÷ ⁻40
    5

24. 11 × ⁻11
    ⁻121

ALGEBRA Use mental math to find the value of y.

25. $y \times {}^-4 = {}^-16$
    $y = 4$

26. $y \div {}^-8 = 5$
    $y = {}^-40$

27. ${}^-6 \times y = 60$
    $y = {}^-10$

28. ${}^-21 \div y = {}^-7$
    $y = 3$

29. $y \times {}^-12 = 12$
    $y = {}^-1$

30. $y \div {}^-3 = {}^-9$
    $y = 27$

Mixed Review

Find the sum or difference.

31. ⁻2 + ⁻13
    ⁻15

32. ⁻16 − ⁻2
    ⁻14

33. ⁻3 − 24
    ⁻27

34. ⁻5 + 10
    5

Write the mixed number as a fraction.

35. $3\frac{2}{5}$
    $\frac{17}{5}$

36. $5\frac{2}{9}$
    $\frac{47}{9}$

37. $1\frac{8}{11}$
    $\frac{19}{11}$

38. $9\frac{3}{8}$
    $\frac{75}{8}$

39. $4\frac{1}{4}$
    $\frac{17}{4}$

Practice **PW57**

Name _____

LESSON 12.6

Explore Operations with Rational Numbers

Find the sum or difference. Estimate to check.

1. $^-4.1 + 6$
   _1.9_

2. $8\frac{1}{5} - {}^-3\frac{1}{2}$
   $11\frac{7}{10}$

3. $^-1\frac{7}{10} + {}^-2\frac{3}{5}$
   $^-4\frac{3}{10}$

4. $6.7 - {}^-2.6$
   _9.3_

5. $^-1\frac{5}{6} + 2\frac{2}{3}$
   $\frac{5}{6}$

6. $12.7 + {}^-3.1$
   _9.6_

7. $^-8.4 - {}^-4.8$
   _$^-3.6$_

8. $^-2\frac{4}{5} - 3\frac{3}{5}$
   $^-6\frac{2}{5}$

Find the product or quotient. Estimate to check.

9. $^-1\frac{3}{8} \div \frac{^-3}{4}$
   $1\frac{5}{6}$

10. $^-4.4 \times 3.3$
    _$^-14.52$_

11. $4\frac{1}{2} \times \frac{^-5}{6}$
    $^-3\frac{3}{4}$

12. $^-0.8 \times {}^-1.7$
    _1.36_

13. $^-2\frac{1}{2} \div \frac{5}{8}$
    _$^-4$_

14. $9.4 \div {}^-5$
    _$^-1.88$_

15. $^-1\frac{4}{5} \times {}^-2\frac{2}{3}$
    $4\frac{4}{5}$

16. $3.2 \div {}^-2.5$
    _$^-1.28$_

Evaluate the expression.

17. $2^3 - ({}^-1\frac{1}{3} + 4)$
    $5\frac{1}{3}$, or $\frac{16}{3}$

18. $9.5 + ({}^-1.8 \times 0.2)$
    _9.14_

19. $^-4 - {}^-2 + (\frac{1}{2} \times 6)$
    _1_

ALGEBRA Evaluate the expression for $x = {}^-1.6$.

20. $^-1.4 + (x - 0.5)$
    _$^-3.5$_

21. $^-12.5 + x$
    _$^-14.1$_

22. $x + 3.8$
    _2.2_

Mixed Review

Find the greatest common factor.

23. 42, 60
    _6_

24. 20, 36
    _4_

25. 55, 99
    _11_

26. 48, 84
    _12_

27. 95, 133
    _19_

Find the mean.

28. 12, 15, 18, 14, 20, 13, 17, 11
    _15_

29. 50, 72, 67, 55, 75, 61, 66, 58
    _63_

30. 94, 78, 90, 83, 88, 95, 96, 80
    _88_

31. 10, 79, 19, 56, 34, 89, 62, 27
    _47_

PW58 Practice

Name _____

LESSON 13.1

Write Expressions

Write an algebraic expression for the word expression.

1. 47 less than the product of y and 7

   $7y - 47$

2. $\frac{3}{4}$ added to 8 times w

   $8w + \frac{3}{4}$

3. q times 11, minus the product of 6 and t

   $11q - 6t$

4. the difference between a and 4, divided by the sum of b and 9

   $(a - 4) \div (b + 9)$

5. the product of m and 15 divided by the sum of n and 50

   $15m \div (n + 50)$

6. k times 12 divided by the product of d and 3

   $12k \div 3d$

Write a word expression for each. Possible expressions are given.

7. $15 - (r + s)$

   fifteen decreased by the sum of r and s

8. $\frac{g + 5}{t}$

   the sum of g and 5, divided by t

9. $k \times m + 1.5$

   1.5 more than the product of k and m

10. $\frac{20}{bc}$

    twenty divided by the product of b and c

11. $7.5n + xy$

    the sum of 7.5 times n and x times y

12. $de - \frac{1}{2}$

    one-half less than the product of d and e

Mixed Review

Find the quotient. Write the answer in simplest form.

13. $\frac{2}{3} \div \frac{1}{4}$

    $2\frac{2}{3}$

14. $\frac{4}{5} \div \frac{2}{15}$

    6

15. $9 \div \frac{3}{8}$

    24

16. $\frac{3}{7} \div \frac{2}{5}$

    $1\frac{1}{14}$

Find the difference.

17. $7 - {}^-10$

    17

18. ${}^-6 - 5$

    ${}^-11$

19. ${}^-12 - {}^-3$

    ${}^-9$

20. $14 - {}^-18$

    32

Name _____

LESSON 13.2

Evaluate Expressions

Evaluate the expression for $x = {}^-5, {}^-1, 0,$ and 3.

1. $4x - 2$
   ${}^-22, {}^-6, {}^-2, 10$

2. $13 - 2x$
   $23, 15, 13, 7$

3. ${}^-7 + 5x$
   ${}^-32, {}^-12, {}^-7, 8$

4. $\frac{3}{4} - 3x$
   $15\frac{3}{4}, 3\frac{3}{4}, \frac{3}{4}, {}^-8\frac{1}{4}$

5. $6 + 10 \cdot (x - 3)$
   ${}^-74, {}^-34, {}^-24, 6$

6. $\frac{12}{x-1} - 8$
   ${}^-10, {}^-14, {}^-20, {}^-2$

7. $25 - x^2$
   $0, 24, 25, 16$

8. $6x \div {}^-3$
   $10, 2, 0, {}^-6$

9. ${}^-4 \cdot (x + 5)$
   $0, {}^-16, {}^-20, {}^-32$

Simplify the expression.
Then evaluate the expression for the given value of the variable.

10. $4x - x + 21$ for $x = 5$
    $3x + 21; 36$

11. $k - 7k - 11$ for $k = {}^-3$
    ${}^-6k - 11; 7$

12. $6a + 3b + 27 - 2a$
    for $a = {}^-7$ and $b = 6$
    $4a + 3b + 27; 17$

13. $m + 30 - 2n + 4m$
    for $m = {}^-6$ and $n = 15$
    $5m - 2n + 30; {}^-30$

Evaluate the expression for the given values of the variables.

14. $4f \cdot (g - h)$
    for $f = {}^-2, g = {}^-10,$ and $h = 12$
    176

15. $r \cdot (6s + 2t)$
    for $r = 4, s = 5,$ and $t = {}^-9$
    48

Mixed Review

Compare. Write $<$ or $>$.

16. ${}^-1.50 \;\underline{>}\; {}^-1.55$

17. $\frac{-2}{3} \;\underline{<}\; \frac{1}{3}$

18. $0.80 \;\underline{>}\; \frac{3}{4}$

19. $\frac{-2}{7} \;\underline{>}\; \frac{-5}{6}$

Find the product.

20. ${}^-20 \times 6$
    ${}^-120$

21. ${}^-12 \times 8$
    ${}^-96$

22. $7 \times {}^-11$
    ${}^-77$

23. ${}^-15 \times 15$
    ${}^-225$

24. ${}^-7 \times {}^-40$
    280

25. ${}^-35 \times 0$
    0

26. $9 \times {}^-12$
    ${}^-108$

27. ${}^-10 \times {}^-11$
    110

Name _____

LESSON 13.4

Expressions with Squares and Square Roots

Evaluate the expression.

1. $\sqrt{16} + 9$

   _13_

2. $34 - \sqrt{36}$

   _28_

3. $11 + \sqrt{49} - 3$

   _15_

4. $64 - \sqrt{64}$

   _56_

5. $2^2 + 10 + \sqrt{25}$

   _19_

6. $\sqrt{16} \times 3$

   _12_

7. $\sqrt{64} \div 8 \times 1$

   _1_

8. $9^2 \div 9 + 9$

   _18_

9. $12^2 \div 6 \times 3$

   _72_

10. $51 - \sqrt{64} \times 6$

    _-3_

11. $(12 + \sqrt{4}) - 14$

    _0_

12. $5 \times (7 - 2^2)$

    _15_

13. $\sqrt{121} + 3 \times 5^2$

    _86_

14. $^-6(\sqrt{81} - \sqrt{64})$

    _-6_

15. $196 \div \sqrt{4} \times 2$

    _196_

Evaluate the expression for the given value of the variable.

16. $x^2 + \sqrt{64}$ for $x = 6$

    _44_

17. $\sqrt{121} - \sqrt{m} + 5$ for $m = 100$

    _6_

18. $(h + 3) - 55$ for $h = \sqrt{49}$

    _-45_

19. $\sqrt{4} \times y^2 + 3$ for $y = 5$

    _53_

20. $(r^2 + \sqrt{16}) \div 2$ for $r = 8$

    _34_

21. $7a^2 - \sqrt{a}$ for $a = 4$

    _110_

Mixed Review

Find the product. Write the answer in simplest form.

22. $\frac{3}{5} \times \frac{2}{3}$

    $\frac{2}{5}$

23. $\frac{5}{8} \times \frac{2}{5}$

    $\frac{1}{4}$

24. $\frac{1}{2} \times \frac{3}{4}$

    $\frac{3}{8}$

25. $\frac{2}{3} \times \frac{1}{6}$

    $\frac{1}{9}$

Write each rational number in the form $\frac{a}{b}$. Possible answers are given.

26. $1\frac{3}{4}$

    $\frac{7}{4}$

27. 0.7

    $\frac{7}{10}$

28. 0.75

    $\frac{75}{100}$

29. 2.25

    $\frac{225}{100}$

Practice PW61

Name _____

LESSON 14.1

Connect Words and Equations

Write an equation for the word sentence. Choice of variable may vary.

1. 12 less than a number equals 15.
   $n - 12 = 15$

2. The quotient of a number and 7 is 63.
   $n \div 7 = 63$ or $\frac{n}{7} = 63$

3. 5 more than a number is 31.
   $n + 5 = 31$

4. 6 less than a number r is 16.
   $r - 6 = 16$

5. 3 times the price p equals $9.45.
   $3p = \$9.45$

6. 4 times the number of cars is 84.
   $4c = 84$

7. A number x divided by 2.5 is 3.5.
   $x \div 2.5 = 3.5$ or $\frac{x}{2.5} = 3.5$

8. 12 fewer than a number m is $17\frac{1}{2}$.
   $m - 12 = 17\frac{1}{2}$

9. 5 times the number of students in the class is 155.
   $5n = 155$

10. The number of auditorium seats divided by 3 is 174.
    $x \div 3 = 174$ or $\frac{x}{3} = 174$

11. Eight more than your test score is 100.
    $t + 8 = 100$

12. The difference between a number k and ⁻7 is 12.
    $k - {}^-7 = 12$

Mixed Review

Rewrite the problem so that the divisor is a whole number.

13. $9.2 \div 5.4$
    $92 \div 54$

14. $7.3 \div 2.6$
    $73 \div 26$

15. $19.12 \div 3.4$
    $191.2 \div 34$

16. $67.3 \div 0.18$
    $6,730 \div 18$

Write the fraction in simplest form.

17. $\frac{7}{21}$
    $\frac{1}{3}$

18. $\frac{16}{30}$
    $\frac{8}{15}$

19. $\frac{9}{24}$
    $\frac{3}{8}$

20. $\frac{15}{50}$
    $\frac{3}{10}$

21. $\frac{20}{45}$
    $\frac{4}{9}$

22. $\frac{12}{18}$
    $\frac{2}{3}$

23. $\frac{10}{15}$
    $\frac{2}{3}$

24. $\frac{24}{36}$
    $\frac{2}{3}$

25. $\frac{33}{55}$
    $\frac{3}{5}$

26. $\frac{16}{24}$
    $\frac{2}{3}$

PW62 Practice

Name _____

LESSON 14.3

Solve Addition Equations

Solve and check.

1. $x + 9 = 14$
   _____ $x = 5$ _____

2. $m + 3.5 = 9$
   _____ $m = 5.5$ _____

3. $12 + w = 23$
   _____ $w = 11$ _____

4. $t + 8.7 = 16.3$
   _____ $t = 7.6$ _____

5. $b + 4\frac{1}{3} = 11$
   _____ $b = 6\frac{2}{3}$ _____

6. $15 = e + 11.2$
   _____ $e = 3.8$ _____

7. $n + 6\frac{3}{5} = 9$
   _____ $n = 2\frac{2}{5}$ _____

8. $18.9 + c = 31.2$
   _____ $c = 12.3$ _____

9. $24.6 = 15.7 + h$
   _____ $h = 8.9$ _____

10. $14\frac{1}{2} + d = 22$
    _____ $d = 7\frac{1}{2}$ _____

11. $5\frac{1}{4} = 2\frac{1}{2} + z$
    _____ $z = 2\frac{3}{4}$ _____

12. $k + 17.8 = 42.1$
    _____ $k = 24.3$ _____

13. $9.3 = 5.9 + q$
    _____ $q = 3.4$ _____

14. $51 = 29.8 + p$
    _____ $p = 21.2$ _____

15. $j + 4 = {}^-7$
    _____ $j = {}^-11$ _____

16. $8.6 + s = 14.3$
    _____ $s = 5.7$ _____

17. ${}^-3 = y + 6$
    _____ $y = {}^-9$ _____

18. $17\frac{3}{5} = a + 8\frac{1}{2}$
    _____ $a = 9\frac{1}{10}$ _____

Mixed Review

Estimate. Possible estimates are given.

19. $67.9 - 39.6$
    _____ 30 _____

20. $109.4 \div 22$
    _____ 5 _____

21. $\$7.78 + \$6.19$
    _____ $14 _____

22. 1.9×15.1
    _____ 30 _____

23. $3 \times \$51.99$
    _____ $150 _____

24. $202.1 - 58.3$
    _____ 140 _____

25. $6.71 + 19.03$
    _____ 26 _____

26. $599.2 \div 3.9$
    _____ 150 _____

Write the corresponding decimal or percent.

27. 16% 0.16
28. 7% 0.07
29. 0.65 65%
30. 19% 0.19
31. 0.54 54%
32. 0.02 2%
33. 10% 0.1 or 0.10
34. 0.42 42%
35. 90% 0.90 or 0.9
36. 0.09 9%

Practice **PW63**

Name _____

LESSON 14.4

Solve Subtraction Equations

Solve and check.

1. $t - 1 = 9$
   $t = 10$

2. $12 = x - 3$
   $x = 15$

3. $b - 6 = 2$
   $b = 8$

4. $4 = a - 3$
   $a = 7$

5. $y - 4 = 19$
   $y = 23$

6. $1 = n - 50$
   $n = 51$

7. $c - 1.5 = 7$
   $c = 8.5$

8. $4.4 = h - 13.4$
   $h = 17.8$

9. $k - 7.3 = 12.7$
   $k = 20$

10. $4\frac{1}{3} = z - \frac{2}{3}$
    $z = 5$

11. $f - 8\frac{3}{4} = 5$
    $f = 13\frac{3}{4}$

12. $10\frac{5}{8} = w - 8$
    $w = 18\frac{5}{8}$

13. $36.5 = g - 18.6$
    $g = 55.1$

14. $e - 2\frac{1}{3} = 4\frac{1}{2}$
    $e = 6\frac{5}{6}$

15. $42 = v - 3\frac{2}{9}$
    $v = 45\frac{2}{9}$

16. $m - 31 = 2\frac{1}{4}$
    $m = 33\frac{1}{4}$

17. $6.8 = p - 14.5$
    $p = 21.3$

18. $s - 1.9 = 5.4$
    $s = 7.3$

Mixed Review

Find the product.

19. $10 \times {}^-5$
    $^-50$

20. $^-9 \times {}^-9$
    81

21. $^-3 \times 12$
    $^-36$

22. $^-11 \times {}^-4$
    44

Find the sum or difference. Write the answer in simplest form.

23. $4\frac{2}{3} + 7\frac{3}{4}$
    $12\frac{5}{12}$

24. $8\frac{5}{8} - 1\frac{2}{5}$
    $7\frac{9}{40}$

25. $2\frac{5}{6} + 3\frac{1}{3}$
    $6\frac{1}{6}$

26. $3\frac{1}{8} - 1\frac{3}{4}$
    $1\frac{3}{8}$

27. $1\frac{3}{4} - 1\frac{1}{6}$
    $\frac{7}{12}$

28. $3\frac{1}{2} + 4\frac{3}{5}$
    $8\frac{1}{10}$

29. $5\frac{5}{7} - 1\frac{1}{2}$
    $4\frac{3}{14}$

30. $5\frac{1}{3} - 2\frac{5}{6}$
    $2\frac{1}{2}$

PW64 Practice

Name _____

LESSON 15.2

Solve Multiplication and Division Equations

Solve and check.

1. $3y = 9$
   _y = 3_

2. $4p = 24$
   _p = 6_

3. $\frac{x}{2} = 7$
   _x = 14_

4. $\frac{s}{3} = 5$
   _s = 15_

5. $20 = 4n$
   _n = 5_

6. $32 = 8k$
   _k = 4_

7. $7 = \frac{a}{9}$
   _a = 63_

8. $4 = \frac{m}{8}$
   _m = 32_

9. $2x = 8$
   _x = 4_

10. $3c = 18$
    _c = 6_

11. $\frac{a}{4} = 8$
    _a = 32_

12. $\frac{m}{5} = 4$
    _m = 20_

13. $6 = \frac{k}{4}$
    _k = 24_

14. $60 = 5y$
    _y = 12_

15. $11 = \frac{b}{6}$
    _b = 66_

16. $45 = 3n$
    _n = 15_

17. $140 = 14g$
    _g = 10_

18. $513 = \frac{w}{3}$
    _w = 1,539_

19. $1{,}320 = 22d$
    _d = 60_

20. $19 = \frac{g}{11}$
    _g = 209_

Solve and check.

21. $\frac{12}{17} = \frac{k}{68}$
    _k = 48_

22. $\frac{x}{7} = 9\frac{1}{3}$
    x = 65$\frac{1}{3}$

23. $\frac{5}{8}n = 3\frac{3}{4}$
    _n = 6_

24. $\frac{2}{3}m = 2\frac{1}{6}$
    m = 3$\frac{1}{4}$

Mixed Review

Solve and check.

25. $a - 11 = 27$
    _a = 38_

26. $18.4 = b - 3.69$
    _b = 22.09_

27. $c - 6\frac{1}{3} = 14\frac{2}{3}$
    _c = 21_

28. $1\frac{5}{8} + n = 5\frac{1}{4}$
    n = 3$\frac{5}{8}$

Multiply. Write the answer in simplest form.

29. $2\frac{1}{4} \times 3\frac{2}{3}$
    8$\frac{1}{4}$

30. $9\frac{1}{5} \times 1\frac{3}{4}$
    16$\frac{1}{10}$

31. $4\frac{3}{8} \times 2\frac{2}{5}$
    10$\frac{1}{2}$

32. $2\frac{2}{5} \times 1\frac{2}{3}$
    _4_

33. $6\frac{3}{5} \times 2\frac{1}{3}$
    15$\frac{2}{5}$

34. $2\frac{2}{3} \times 3\frac{3}{8}$
    _9_

35. $1\frac{1}{2} \times 5\frac{5}{6}$
    8$\frac{3}{4}$

36. $1\frac{1}{4} \times 3\frac{1}{5}$
    _4_

Practice

Name _____

LESSON 15.3

Use Formulas

Use the formula $d = rt$ to complete.

1. $d =$ __80 mi__
   $r = 20$ mi per hr
   $t = 4$ hr

2. $d =$ __714 ft__
   $r = 17$ ft per sec
   $t = 42$ sec

3. $d =$ __51.94 km__
   $r = 9.8$ km per hr
   $t = 5.3$ hr

4. $d = 75$ mi
   $r =$ __25 mi per hr__
   $t = 3$ hr

5. $d = 1{,}320$ km
   $r =$ __6 km per min__
   $t = 220$ min

6. $d = 99$ ft
   $r =$ __9 ft per sec__
   $t = 11$ sec

7. $d = 605$ mi
   $r = 55$ mi per hr
   $t =$ __11 hr__

8. $d = 336$ ft
   $r = 28$ ft per sec
   $t =$ __12 sec__

9. $d = 500$ ft
   $r = 25$ ft per min
   $t =$ __20 min__

Convert the temperature to degrees Fahrenheit. Write your answer as a decimal.

10. 30°C __86°F__
11. 25°C __77°F__
12. 50°C __122°F__
13. 13°C __55.4°F__
14. 3°C __37.4°F__
15. 60°C __140°F__

16. 22°C __71.6°F__
17. 54°C __129.2°F__
18. 7°C __44.6°F__
19. 100°C __212°F__
20. 15°C __59°F__
21. 0°C __32°F__

Convert the temperature to degrees Celsius. Write your answer as a decimal and round to the nearest tenth of a degree.

22. 71°F __21.7°C__
23. 50°F __10°C__
24. 140°F __60°C__
25. 90°F __32.2°C__
26. 45°F __7.2°C__
27. 121°F __49.4°C__

28. 32°F __0°C__
29. 49°F __9.4°C__
30. 96°F __35.6°C__
31. 130°F __54.4°C__
32. 113°F __45°C__
33. 86°F __30°C__

Mixed Review

Write each rational number in the form $\frac{a}{b}$. Possible answers are given.

34. $2\frac{1}{3}$ __$\frac{7}{3}$__
35. 5.1 __$\frac{51}{10}$__
36. $^-8\frac{2}{5}$ __$\frac{-42}{5}$__
37. $^-1.667$ __$\frac{-1{,}667}{1{,}000}$__

Estimate. Possible estimates are given.

38. 5.4×19.7 __100__
39. $41.6 \div 6.8$ __6__
40. $187.51 - 90.4$ __100__
41. $276.7 + 389.5$ __700__

PW66 Practice

Name _____

LESSON 15.5

Problem Solving Strategy: Work Backward

Solve the problem by working backward.

1. Keesha went to the movies with her brother, Merle, and spent $15.00. The tickets cost $4.50 each. She bought a box of popcorn and 2 drinks. The drinks cost $1.50 each. How much did the popcorn cost?

   _____ $3.00 _____

2. Alex brought 48 cookies to school to celebrate his birthday. He gave 9 to teachers. He then shared equally the remaining cookies with his 18 classmates. How many cookies remained?

   _____ 1 cookie _____

3. An engineer is checking wells on a hillside. He starts at his van and walks up 100 m to Well 1. He climbs down 50 m to Well 2. Then he climbs up 200 m to Well 3, which is 220 m above Well 4. How high is each well from the engineer's van?

   _____ Well 1, 100 m; Well 2, 50 m; _____
   _____ Well 3, 250 m; Well 4, 30 m _____

4. Karen had a bag of oats. She used $1\frac{1}{4}$ c in a meatloaf and $3\frac{1}{4}$ c to make cookies. To make granola, Karen used twice the amount of oats she used to make cookies. If there are 4 c of oats left over, how much did Karen start with?

   _____ 15 c _____

5. Maya paid $174 for a car she rented for 4 days. The rate was $36 per day. Maya also had to pay $0.20 per mi after the first 200 mi driven. How many miles did Maya drive the rented car?

   _____ 350 mi _____

6. Miguel poured some punch into the pitcher. Tim added 16 oz more. Bill then added enough punch to double the amount in the pitcher. The pitcher now contains 72 oz of punch. How much did Miguel pour into the pitcher?

   _____ 20 oz _____

Mixed Review

Write an algebraic expression for the word expression.

7. 4.7 more than 5 times x

   _____ $5x + 4.7$ _____

8. 5 less than the quotient of t and 4.2

   _____ $t \div 4.2 - 5$ _____

9. the product of p, $4n$ and m

   _____ $p \times 4n \times m$ _____

Find the product.

10. $^-9 \times 5$ _____ $^-45$ _____

11. $15 \times ^-3$ _____ $^-45$ _____

12. $^-8 \times ^-6$ _____ 48 _____

13. $22 \times ^-7$ _____ $^-154$ _____

14. $^-12 \times ^-5$ _____ 60 _____

15. $^-105 \times 3$ _____ $^-315$ _____

Practice **PW67**

Name _____

LESSON 16.1

Points, Lines, and Planes

Name the geometric figure.

1.

   line segment XY

2.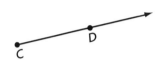

   ray CD

3. • A

   point A

4.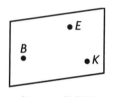

   plane BEK

5.

   line MN

6.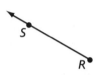

   ray RS

For Exercises 7–9, use the figure at the right.

7. Name three points.

   point A, point B, point C

8. Name four different rays.

   _AB or AC, CB or CA, BC, BA_

9. Name three different line segments.

   _AB or BA, AC or CA, BC or CB_

Mixed Review

Solve and check.

10. $\frac{n}{5} = 40$

    _n = 200_

11. $7x = 63$

    _x = 9_

12. $^-25 = \frac{k}{6}$

    _k = $^-$150_

13. $4.7 = \frac{d}{2.1}$

    _d = 9.87_

14. $84 = 7c$

    _c = 12_

15. $s \div 1.05 = 800$

    _s = 840_

16. $^-3m = ^-450$

    _m = 150_

17. $129.5 = ^-7g$

    _g = $^-$18.5_

Write the sum or difference. Write the answer in simplest form.

18. $\frac{2}{3} + \frac{1}{4}$

    $\frac{11}{12}$

19. $\frac{3}{5} - \frac{1}{2}$

    $\frac{1}{10}$

20. $\frac{1}{8} + \frac{1}{6}$

    $\frac{7}{24}$

21. $\frac{5}{8} - \frac{2}{5}$

    $\frac{9}{40}$

22. $\frac{1}{3} - \frac{2}{9}$

    $\frac{1}{9}$

23. $\frac{1}{4} + \frac{3}{5}$

    $\frac{17}{20}$

24. $\frac{3}{4} - \frac{1}{3}$

    $\frac{5}{12}$

25. $\frac{3}{7} + \frac{1}{2}$

    $\frac{13}{14}$

PW68 Practice

Name _____

LESSON 16.3

Angle Relationships

For 1–4 use the figure.

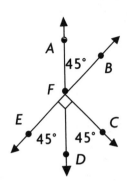

1. Name two angles adjacent to ∠AFB.

   __∠AFE; ∠BFC__

2. Name an angle vertical to ∠EFD.

   __∠AFB__

3. Name an angle that is complementary to ∠DFC.

   __∠DFE__

4. Name two angles that are supplementary to ∠AFE.

   __∠AFB; ∠EFD__

Find the unknown angle measure. The angles are complementary or supplementary.

5.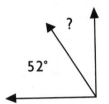

   __38°__

6.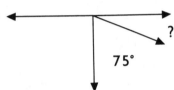

   __15°__

7.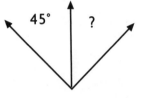

   __45°__

8.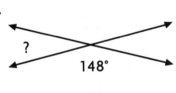

   __32°__

9.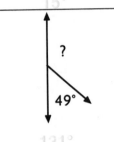

   __131°__

10.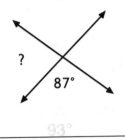

    __93°__

Mixed Review

Solve and check.

11. $y + 9 = 14$

    __y = 5__

12. $5 + c = {}^-7$

    __c = ⁻12__

13. $4.3 = x + 1.8$

    __x = 2.5__

14. $p + 9\frac{1}{3} = 14\frac{2}{3}$

    __p = 5\frac{1}{3}__

Write the prime factorization in exponent form.

15. 24

    __$2^3 \times 3$__

16. 144

    __$2^4 \times 3^2$__

17. 360

    __$2^3 \times 3^2 \times 5$__

Practice PW69

Name _____

LESSON 16.4

Classify Lines

Classify the lines.

1.
2.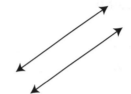
3.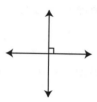

   ____intersecting____   ____parallel____   ____perpendicular, intersecting____

The lines in the figure at the right intersect to form a cube. All possible answers are given.

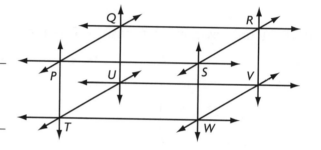

4. Name all lines that are parallel to $\overleftrightarrow{TW}$.

   ____UV, PS, QR____

5. Name all the lines that intersect $\overleftrightarrow{PQ}$.

   ____PS, QR, PT, QU____

6. Name all the lines that are perpendicular to and intersect $\overleftrightarrow{UQ}$.

   ____PQ, QR, TU, UV____

7. Name all the lines that are parallel to $\overleftrightarrow{RV}$. ____SW, PT, QU____

8. Is line $\overleftrightarrow{PS}$ perpendicular to line $\overleftrightarrow{QU}$? ____no____

9. Is line $\overleftrightarrow{PQ}$ parallel to line $\overleftrightarrow{UV}$? ____no____

Mixed Review

Find the product.

10. $3 \times {}^-9$ ____ ⁻27 ____   11. $^-12 \times 8$ ____ ⁻96 ____   12. $^-4 \times {}^-10$ ____ 40 ____   13. $^-6 \times 15$ ____ ⁻90 ____

14. $^-11 \times {}^-10$ ____ 110 ____   15. $^-5 \times 20$ ____ ⁻100 ____   16. $^-9 \times 8$ ____ ⁻72 ____   17. $7 \times {}^-50$ ____ ⁻350 ____

18. $12.29 - 1.07$ ____ 11.22 ____   19. $8.791 + 0.45$ ____ 9.241 ____

20. $0.602 - 0.060$ ____ 0.542 ____   21. $527.4 + 43.685$ ____ 571.085 ____

PW70 Practice

Name _____

LESSON 17.1

Triangles

Find the measure of the missing angle and classify the triangle containing the angle.

1.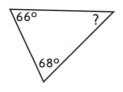

   __46°; acute__

2.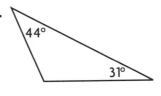

   __105°; obtuse__

3.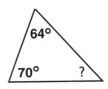

   __46°; acute__

4.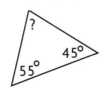

   __80°; acute__

5.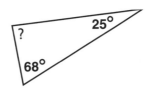

   __87°; right__

6.

   __70°; right__

7.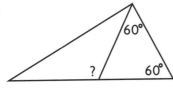

   __120°; obtuse__

8.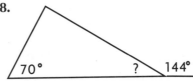

   __36°; acute__

9.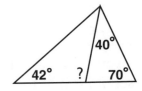

   __110°; obtuse__

For 10–17, use the figure at the right. Line AB is parallel to line CD. Find the measure of each angle.

10. ∠1 __60°__
11. ∠2 __50°__
12. ∠3 __110°__
13. ∠4 __130°__
14. ∠5 __70°__
15. ∠6 __110°__
16. ∠7 __130°__
17. ∠8 __50°__

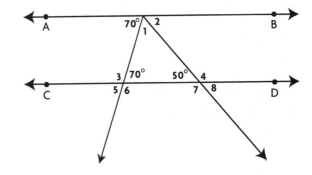

Mixed Review

Evaluate.

18. $^+4 - {^-6} - {^+10}$ __0__

19. $^-3 - {^+2} - {^-4}$ __⁻1__

20. $^-8 - {^-5} - {^+7}$ __⁻10__

Evaluate the expression.

21. $^-5(\sqrt{64} - 3)$ __⁻25__

22. $3^2 + \sqrt{81}$ __18__

23. $\sqrt{100} \div 5 \times 2^2$ __8__

Practice PW71

Name _____

LESSON 17.2

Problem Solving Strategy: Find a Pattern

Solve the problem by finding a pattern.

1. The Auto Stop is advertising a special sale: buy 3 cans of motor oil, get 1 can free. How many cans should you buy in order to get 36 cans of motor oil?

   _____ 27 cans _____

2. The Auto Stop charges $2.09 for a can of motor oil. Adam spends $37.62 on oil during the "buy 3 cans, get 1 free" sale. How many cans of oil did he get in all?

   _____ 24 cans _____

3. The school cafeteria serves both ice cream and apples for dessert. Twenty-five students choose ice cream for every 6 students who choose apples. In one week, the cafeteria served 600 ice creams. How many students chose apples?

   _____ 144 students _____

4. Barry and Cecilia are playing a number game. One of them thinks of a number pattern and gives the first six numbers. The other has to name the next number in the pattern. Barry gave Cecilia these numbers: 3, 4, 7, 11, 18, 29. Cecilia correctly gave the next number. What number did she give?

   _____ 47 _____

5. Twenty-four students went on the school trip to the science museum. The admission price was $4.00 per student, but 1 student was admitted free for every 3 students who paid. What was the total cost?

   _____ $72.00 _____

6. The floor of a 17 ft by 13 ft sun room is tiled with tiles that are 1 ft². The tiles alternate between black and white. If there is a black tile in one corner of the room, how many black tiles will be needed in all?

   _____ 111 black tiles _____

Mixed Review

Use a property to simplify the expression. Then evaluate the expression and identify the property you used.

7. ⁻3 + 16 + 23

   _____ 36; commutative _____

8. (24 + 37) + 63

   _____ 124; associative _____

9. ⁻73 + 120 + ⁻27

   _____ 20; commutative _____

Write the fraction in simplest form.

10. $\frac{14}{48}$ 11. $\frac{27}{45}$ 12. $\frac{24}{60}$ 13. $\frac{15}{90}$ 14. $\frac{20}{64}$

    $\frac{7}{24}$ $\frac{3}{5}$ $\frac{2}{5}$ $\frac{1}{6}$ $\frac{5}{16}$

Name _____

LESSON 17.3

Quadrilaterals

Name the geometric figure.

1.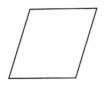

   _____rhombus_____

2.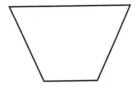

   _____trapezoid_____

3.

   _____quadrilateral_____

4.

   _____square_____

5.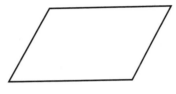

   _____parallelogram_____

6.

   _____rectangle_____

Complete the statement, giving the most exact name for the figure.

7. A quadrilateral with exactly one pair of parallel sides is a

   _____trapezoid_____

8. A polygon with four sides and no pair of parallel sides is a

   _____quadrilateral_____

Diagonals of a quadrilateral are lines drawn from one vertex to the opposite vertex. Complete the statements about diagonals.

9. If a quadrilateral has four congruent sides, but its diagonals are not congruent, then the quadrilateral is a

   _____rhombus_____

10. If a quadrilateral has four congruent sides and its diagonals are congruent, then the quadrilateral is a

    _____square_____

Mixed Review

Evaluate each expression.

11. $3.81 \div m$ for $m = 3$

    _____1.27_____

12. $9w$ for $w = 4.7$

    _____42.3_____

13. $8.02 - r$ for $r = 5.6$

    _____2.42_____

Tell whether you would survey the population or use a sample. Explain.

14. You want to know how far each student in your class lives from school.

    _____survey; small group_____

15. You want to know the percentage of a certain model of car that is red.

    _____sample; large group_____

Practice PW73

Name _____

LESSON 17.4

Draw Two-Dimensional Figures

Draw the figure. Use square dot paper or isometric dot paper. Check students' drawings.

1. an obtuse isosceles triangle

2. a quadrilateral with opposite sides congruent and no right angles

3. a quadrilateral with exactly one pair of parallel sides

4. a pentagon with three congruent sides

5. a quadrilateral with no congruent sides

6. a triangle with all sides congruent

7. a pentagon with all sides congruent

8. a quadrilateral with four right angles and two pairs of congruent sides

9. a quadrilateral with all sides congruent and four right angles

10. a hexagon with all sides congruent

11. a rectangle with all sides congruent

12. a right scalene triangle

Mixed Review

Solve and check.

13. $8.7 = 5.8 + w$

14. $y + 3.6 = 17.1$

15. $23.5 + c = 35.3$

_____ $w = 2.9$ _____

_____ $y = 13.5$ _____

_____ $c = 11.8$ _____

Compare the fractions. Write $<$, $>$, or $=$ for each.

16. $\frac{1}{3}$ $<$ $\frac{5}{9}$

17. $\frac{2}{5}$ $>$ $\frac{3}{10}$

18. $\frac{5}{8}$ $<$ $\frac{3}{4}$

19. $\frac{6}{10}$ $=$ $\frac{3}{5}$

PW74 Practice

Name _____

LESSON 17.5

Circles

For 1–6 use the circle at the right. Name the given parts of the circle.

1. center _____O_____
2. diameters _____DE_____
3. radii _____OC, OD, OE_____
4. four arcs Possible answers: _____CD, DH, HJ, JE_____
5. chords other than diameters _____HJ_____
6. How many sectors are shown in the circle? _____3 sectors_____

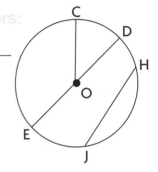

For 7–12 complete the sentence by using *must*, *can*, or *cannot*.

7. Two radii of the same circle ___must___ be equal in length.

8. An arc ___cannot___ pass through the center of a circle.

9. Two chords of the same circle ___can___ be equal in length.

10. A chord drawn through the center of a circle ___must___ be the longest line segment that can be drawn in the circle.

11. Two sectors of a circle ___can___ be equal in area.

12. As the size of a circle increases, the relationship between the radius and the diameter ___cannot___ change.

Mixed Review

Evaluate the expression.

13. $^-2 + 6^2 - 3 + 9$ _____40_____

14. $9 \div 3 \times 4 + (10 - 6)$ _____16_____

15. $2^3 + 4 \times {}^-5 - 1$ _____ $^-13$ _____

16. $(4 \times 6) - ({}^-8 \times 3)$ _____48_____

Find the measure of each angle.

17. The complement of the angle is 18°.

_____72°_____

18. The supplement of the angle is 73°

_____107°_____

19. The supplement of the angle is 126°

_____54°_____

Practice PW75

Name _____

LESSON 18.1

Types of Solid Figures

Classify the figure. Is it a polyhedron?

1.    2.   3. 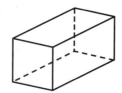   4.

   pentagonal   cone; no   rectangular   cylinder; no
   pyramid; yes   prism; yes

Write *true* or *false* for each statement. Rewrite each false statement as a true statement.

5. A cylinder has one base.

   False; a cylinder has two bases.

6. A cone has one flat surface.

   True

7. A cube has 8 faces

   False; a cube has 6 faces.

8. A square pyramid is a polyhedron.

   True

9. A triangular prism has 2 congruent bases.

   True

10. The faces of a square pyramid are squares.

    False; other than the base, they are triangles.

Mixed Review

Solve and check.

11. $a - 40 = 21$

    $a = 61$

12. $b - 3 = 18$

    $b = 21$

13. $75 = c - 48$

    $c = 123$

14. $^-16 = d - 9$

    $d = ^-7$

15. $14\frac{1}{2} = e - 11\frac{1}{2}$

    $e = 26$

16. $^-7.3 = f - 4$

    $f = ^-3.3$

Write the equal factors. Then find the value.

17. 7^3

    $7 \times 7 \times 7$; 343

18. 9^2

    9×9; 81

19. 4^4

    $4 \times 4 \times 4 \times 4$; 256

PW76 Practice

Name _____

LESSON 18.2

Different Views of Solid Figures

Triangular Pyramid	Triangular Prism	Rectangular Pyramid	Rectangular Prism	Pentagonal Pyramid	Hexagonal Prism	Cylinder	Cone

Name each solid that has the given top view. Refer to the solids in the box above.

1.   2.   3.   4.   5.

   hexagonal   cone   rectangular   pentagonal   triangular
   prism   prism   pyramid   prism

Name the solid figure that has the given views.

6.   7.   8.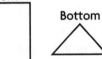

   _____cylinder_____   ___rectangular pyramid___   ___triangular prism___

Mixed Review

9. $\sqrt{100} \times (4 - 3^2) + 9^2$ 10. $(8 - 3)^2 - (\sqrt{49} + \sqrt{4})$

   _____31_____   _____16_____

Write each rational number in the form $\frac{a}{b}$. Possible answers are given.

11. 4.75   12. $6\frac{1}{8}$ 13. $^-6.3$ 14. $10\frac{1}{2}$

    $\frac{475}{100}$ $\frac{49}{8}$ $\frac{-63}{10}$ $\frac{21}{2}$

Name _____

LESSON 18.4

Problem Solving Strategy: Solve a Simpler Problem

Solve by first solving a simpler problem.

1. Jon is building models of edible prisms. He uses gumdrops for vertices and licorice for edges. How many gumdrops and pieces of licorice will he need to make a prism whose base has 8 sides?

   _____16 gumdrops and_____
   _____24 pieces of licorice_____

2. Carol wants to make a model of a prism whose base has 9 sides. She will use balls of clay for the vertices and straws for the edges. How many balls of clay and straws will she need? How many faces will her prism have?

   _____18 balls of clay and 27 straws;_____
   _____11 faces_____

3. Chloe used 30 toothpicks as edges to make a model for a prism. How many sides does one base have? How many vertices does the prism have?

   _____10 sides; 20 vertices_____

4. Dan used 12 balls of clay as vertices to make a model for a prism. How many sides does one base have? How many edges does the prism have?

   _____6 sides; 18 edges_____

5. Marty built a model of a solid figure. It has 6 vertices and 9 edges. It has 5 faces. What is this figure?

   _____triangular prism_____

6. Nancy built a model of a solid figure. It has 5 vertices and 8 edges. It has 5 faces. What is this figure?

   _____square pyramid_____

Mixed Review

For 7–11, use the figure at the right. Find the measure of each angle.

7. ∠BCO _____67°_____

8. ∠BOC _____70°_____

9. ∠COD _____110°_____

10. ∠BOE _____110°_____

11. ∠ODC _____42°_____

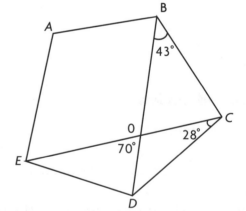

Find the sum or difference.

12. 306 + 1,229 + 558 + 74 _____2,167_____

13. 45,923 + 7,192 + 19,537 _____72,652_____

14. 727,401 − 204,854 _____522,547_____

15. 93,144 − 3,019 _____90,125_____

PW78 Practice

Name _____

LESSON 19.1

Construct Congruent Segments and Angles

Use a compass to decide which pair of line segments in each group is congruent.

1.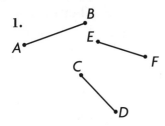

2.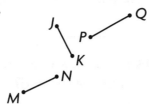

3.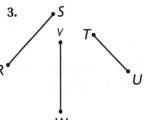

_____CD ≅ EF_____   _____JK ≅ MN_____   _____RS ≅ VW_____

Find the measure of each angle, using a protractor. Then tell whether the angles in each pair are congruent. Write *yes* or *no*.

4.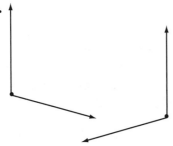

5.

6.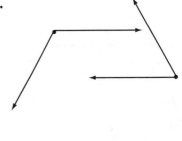

_____105°; yes_____   _____22°; yes_____   _____120°, 60°; no_____

7. In the space at the right, use a compass, and a straightedge to construct a line segment and an angle congruent to the ones below. Check students' drawings.

   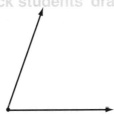

Mixed Review

Write as a decimal.

8. $\frac{3}{5}$ ___0.6___   9. $\frac{7}{10}$ ___0.7___   10. $\frac{1}{4}$ ___0.25___   11. $\frac{5}{2}$ ___2.5___

Solve.

12. $4x = 60$ 13. $\frac{c}{3} = 25$ 14. $36 = 9k$ 15. $3.2 = \frac{s}{7}$

___x = 15___   ___c = 75___   ___k = 4___   ___s = 22.4___

Practice   PW79

Name _____

LESSON 19.2

Bisect Line Segments and Angles

If a line segment of the given length is bisected, how long will each of the smaller segments be?

1. 21 in.   2. 1.08 m   3. 63.35 cm   4. 0.5 in.   5. 13 cm

   10.5 in.   0.54 m   31.675 cm   0.25 in.   6.5 cm

If an angle of the given measure is bisected, how many degrees will there be in each of the smaller angles that are formed?

6. 84°   7. 27.4°   8. 108.5°   9. 12.5°   10. 27°

   42°   13.7°   54.25°   6.25°   13.5°

Bisect the figures.

11.

Check students' drawings.

12.

J _____ K

Check students' drawings.

13.

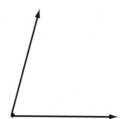

Check students' drawings.

14.

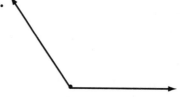

Check students' drawings.

Mixed Review

Solve.

15. $n - 12 = 23$ 16. $g + 13.8 = 23$ 17. $a - 64 = 15$ 18. $2.5 + m = 7$

    $n = 35$ $g = 9.2$ $a = 79$ $m = 4.5$

Solve.

19. $\frac{2}{3} \times \frac{5}{8}$ 20. $\frac{6}{7} \div \frac{1}{3}$ 21. $\frac{4}{5} \times \frac{1}{3}$ 22. $\frac{4}{9} \div \frac{2}{3}$

    $\frac{5}{12}$ $2\frac{4}{7}$ $\frac{4}{15}$ $\frac{2}{3}$

PW80 Practice

Name _____

LESSON 19.4

Similar and Congruent Figures

Tell whether the figures in each pair appear to be *similar, congruent, both,* or *neither.*

1.
   __both__

2.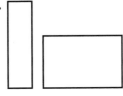
   __neither__

3.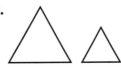
   __similar__

4.
   __both__

5.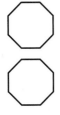
   __both__

6.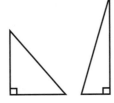
   __neither__

7.
   __similar__

8.
   __neither__

Write *true* or *false* for each statement.

9. If two figures are the same shape and size, then they must be congruent.

   __true__

10. If two figures are similar, then they must be congruent.

    __false__

Mixed Review

Add.

11. $14 + {}^-8$ __6__
12. ${}^-7 + {}^-4$ __$^-$11__
13. ${}^-28 + 19$ __$^-$9__
14. $13 + {}^-15$ __$^-$2__

Evaluate each expression for $a = 3$ and $b = 5$.

15. $a^2 + b^2 - a + (b - 7)$

    __29__

16. $8b - 23 + (ab - 11) - 7a$

    __0__

17. $2a - (6 + 3b) + ab^2$

    __60__

18. $3b + 20 - (\frac{b}{5} + 8a^2)$

    __$^-$38__

Practice PW81

Name _____ LESSON 20.1

Ratios and Rates

Write two equivalent ratios. Possible answers are given.

1. $\frac{4}{6}$ 　　　　　2. $\frac{12}{28}$ 　　　　　3. $\frac{5}{20}$

　　$\frac{2}{3}, \frac{8}{12}$ _____　　　　$\frac{6}{14}, \frac{3}{7}$ _____　　　　$\frac{1}{4}, \frac{10}{40}$ _____

4. $\frac{2}{18}$ 　　　　　5. $\frac{7}{49}$ 　　　　　6. $\frac{2}{5}$

　　$\frac{1}{9}, \frac{3}{27}$ _____　　　　$\frac{1}{7}, \frac{3}{21}$ _____　　　　$\frac{4}{10}, \frac{8}{20}$ _____

Write each ratio in fraction form. Then find the unit rate.

7. 7 apples for $1.00　　8. $0.72 for 12 pages　　9. 24 people in 6 cars

　$\frac{7\ apples}{\$1.00}$; $0.14 per apple 　$\frac{\$0.72}{12\ pages}$; $0.06 per page 　$\frac{24\ people}{6\ cars}$; 4 people per car

10. 65 mi per 3 gal　　11. $49 for 5 CDs　　12. $20 per dozen tarts

　$\frac{65\ mi}{3\ gal}$; 21.67 mi per gal 　$\frac{\$49}{5\ CDs}$; $9.80 per CD 　$\frac{\$20}{12\ tarts}$; $1.67 per tart

For Exercises 13–14, use the figure at the right.

13. Find the ratio of unshaded sections to shaded sections. Then write three equivalent ratios.

　　____2 to 4; possible answers: 1:2, 4:8, 6:12____

14. Find the ratio of shaded sections to all the sections. Then write three equivalent ratios.

　　____4 to 6; possible answers: 2:3, 8:12, 12:18____

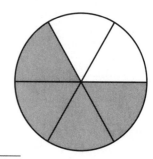

Find the missing term that makes the ratios equivalent.

15. $\frac{3}{7}, \frac{\blacksquare}{14}$ __6__　　16. 7 to 5, ■ to 15 __21__　　17. 15:5, 3:■ __1__

Mixed Review

Find the quotient. Write the answer in simplest form.

18. $\frac{7}{8} \div \frac{3}{4}$ __$1\frac{1}{6}$__　　19. $\frac{2}{3} \div \frac{1}{5}$ __$3\frac{1}{3}$__　　20. $5 \div \frac{1}{4}$ __20__　　21. $2\frac{1}{2} \div \frac{3}{8}$ __$6\frac{2}{3}$__

Compare the fractions. Write $<$, $>$, or $=$ for each $\bigcirc$.

22. $\frac{-1}{3}$ $\bigcirc\!>$ $\frac{-2}{3}$　　23. $\frac{5}{8}$ $\bigcirc\!<$ 0.75　　24. 0.34 $\bigcirc\!>$ $^-1$　　25. 0.25 $\bigcirc\!=$ $\frac{1}{4}$

PW82 Practice

Name _____

LESSON 20.3

Problem Solving Strategy

Write an Equation

Solve the problem by writing an equation.

1. A dripping faucet wastes 3 cups of water in 24 hr. How much water is wasted in 56 hours?

   _____ 7 c _____

2. A map uses the scale of 3 cm for every 10 km. If the map shows a distance of 12 cm, what is the actual distance?

   _____ 40 km _____

3. A pump empties the pool at the rate of 1,000 gal every 4 hours. How long does it take to pump out 20,000 gallons?

   _____ 80 hr _____

4. Tom drinks 8 oz of water for every 3 miles he bikes. After 21 miles, how much water did he drink?

   _____ 56 oz _____

5. A punch consists of 2 parts ginger ale and 3 parts orange juice. If the punch bowl contains 8 c of ginger ale, how many cups of punch are in the bowl?

   _____ 20 c _____

6. A 5-lb bag of apples contains 12 apples. What will a bag of 40 apples weigh?

   _____ $16\frac{2}{3}$ lb _____

7. Jane can iron 4 shirts in 1 hr. At this rate, how long will it take her to iron 10 shirts?

   _____ $2\frac{1}{2}$ hr _____

8. Sol is going on a trip of 275 mi. If he drives a steady 50 mi per hr, how long should the trip take?

   _____ $5\frac{1}{2}$ hr _____

Mixed Review

Find the unknown angle measure.

9.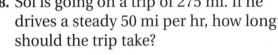

10.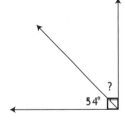

11.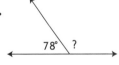

_____ 145° _____ _____ 36° _____ _____ 102° _____

Evaluate the expression for $m = 5$ and $n = 2$.

12. $m - n^2 + 18 \div 3$ _____ 7 _____

13. $30 \div (m^2 - 10) + 6 \times n$ _____ 14 _____

14. $n \times 8(m - 2) - 4^2$ _____ 32 _____

15. $10 - (4 - n) \div (m + 3)$ _____ $9\frac{3}{4}$ _____

Practice PW83

Name _____

LESSON 20.4

Algebra: Ratios and Similar Figures

Name the corresponding sides and angles. Write the ratio of the corresponding sides in simplest form.

1. 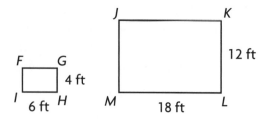 2.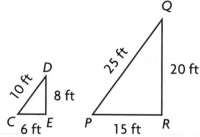

FG corresponds to JK; GH corresponds to KL; HI corresponds to LM; IF corresponds to MJ; ∠F corresponds to ∠J; ∠G corresponds to ∠K; ∠H corresponds to ∠L; ∠I corresponds to ∠M; $\frac{1}{3}$ or $\frac{3}{1}$

CD corresponds to PQ; DE corresponds to QR; EC corresponds to RP; ∠C corresponds to ∠P; ∠D corresponds to ∠Q; ∠E corresponds to ∠R; $\frac{2}{5}$ or $\frac{5}{2}$

Tell whether the figures in each pair are similar. Write *yes* or *no*. If you write *no*, explain.

3. 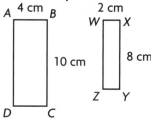 4.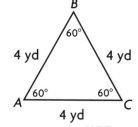

No; ratios are not equivalent.   yes

The figures in each pair are similar. Find the missing measures.

5. 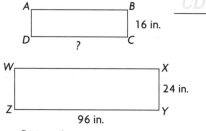 6.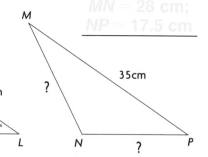

CD = 64 in.   ∠J = 30°; MN = 28 cm; NP = 17.5 cm

Mixed Review

Write the mixed number as a fraction.

7. $7\frac{2}{3}$ $\frac{23}{3}$ 8. $2\frac{8}{9}$ $\frac{26}{9}$ 9. $10\frac{1}{6}$ $\frac{61}{6}$ 10. $5\frac{4}{5}$ $\frac{29}{5}$

Find the difference.

11. $^+12 - ^-7$ 19   12. $^-4 - ^-8$ 4   13. $^-21 - ^+5$ $^-26$ 14. $^-15 - ^-8$ $^-7$

PW84 Practice

Name _____

LESSON 20.5

Algebra: Proportions and Similar Figures

The figures in each pair are similar. Write a proportion. Then find the unknown length.

1.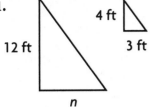

2.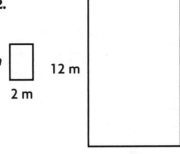

3.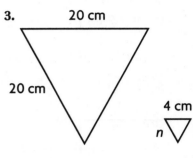

$\frac{3}{n} = \frac{4}{12}$; $n = 9$ ft $\frac{n}{12} = \frac{2}{8}$; $n = 3$ m $\frac{n}{20} = \frac{4}{20}$; $n = 4$ cm

4.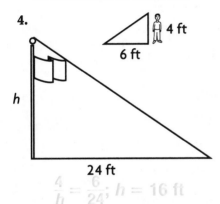

5.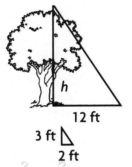

6.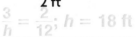

$\frac{4}{h} = \frac{6}{24}$; $h = 16$ ft $\frac{3}{h} = \frac{2}{12}$; $h = 18$ ft $\frac{5}{h} = \frac{10}{320}$; $h = 160$ ft

Mixed Review

Solve and check.

7. $n + 5.5 = 23.1$

8. $4\frac{1}{2} = k + \frac{3}{4}$

9. $28 + w = 104$

   $n = 17.6$ $k = 3\frac{3}{4}$ $w = 76$

Find the LCM of each pair of numbers.

10. 4, 15 __60__   11. 7, 9 __63__   12. 6, 16 __48__   13. 12, 25 __300__

Practice   PW85

Name _____

LESSON 20.6

Algebra: Scale Drawings

Find the unknown dimension.

1. scale: 1 in.:8 ft
   drawing length: 3 in.
   actual length: __24__ ft

2. scale: 1 in.:3 ft
   drawing length: __4__ in.
   actual length: 12 ft

3. scale: 1 cm = 15 km
   drawing length: __9__ cm
   actual length: 135 km

4. scale: 4 cm = 1 mm
   drawing length: 1 cm
   actual length: __0.25__ mm

5. scale: 1 mm:12 m
   drawing length: 9 mm
   actual length: __108__ m

6. scale: 5 in.:35 yd
   drawing length: __1__ in.
   actual length: 7 yd

7. scale: 3 cm = 10 km
   drawing length: __19.5__ cm
   actual length: 65 km

8. scale: 8 cm = 3 mm
   drawing length: 4 cm
   actual length: __1.5__ mm

9. scale: 10 in.:88 yd
   drawing length: 2 in.
   actual length: __$17\frac{3}{5}$__ yd

10. scale: 1 in.:12 ft
    drawing length: __12__ in.
    actual length: 144 ft

11. scale: 1 mm = 25 m
    drawing length: __14__ mm
    actual length: 350 m

12. scale: 1 cm = 3 km
    drawing length: 15 cm
    actual length: __45__ km

Mixed Review

Find the measure of the missing angle and classify the triangle.

13.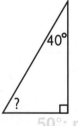
    50°; right

14.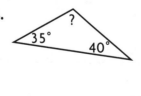
    105°; obtuse

15.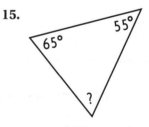
    60°; acute

Simplify the expression. Then evaluate the expression for $x = {}^-5$.

16. $2x + x^2 - 7 - 8x$
    48

17. $5x + 15 + 3x - 2$
    ⁻27

18. $59 + 7x - 6 + 4x$
    ⁻2

PW86 Practice

Name _____

LESSON 20.7

Algebra: Maps

The map distance is given. Find the actual distance. The scale is
1 in. = 20 mi.

1. 4 in.

    _____80 mi_____

2. 20 in.

    _____400 mi_____

3. $1\frac{1}{2}$ in.

    _____30 mi_____

4. 6 in.

    _____120 mi_____

5. 18 in.

    _____360 mi_____

6. $2\frac{1}{2}$ in.

    _____50 mi_____

7. $3\frac{1}{2}$ in.

    _____70 mi_____

8. $5\frac{1}{2}$ in.

    _____110 mi_____

The actual distance is given. Find the map distance. The scale is
1 in. = 20 mi.

9. 250 mi

    _____$12\frac{1}{2}$ in._____

10. 100 mi

    _____5 in._____

11. 150 mi

    _____$7\frac{1}{2}$ in._____

12. 170 mi

    _____$8\frac{1}{2}$ in._____

13. 500 mi

    _____25 in._____

14. 190 mi

    _____$9\frac{1}{2}$ in._____

15. 220 mi

    _____11 in._____

16. 580 mi

    _____29 in._____

Mixed Review

Find the mean, median, and mode.

17. 27, 19, 24, 29, 18, 25
    29, 23, 28

    _____$24\frac{2}{3}$; 25; 29_____

18. 39, 51, 45, 69, 22, 41, 33,
    57, 30

    _____43; 41; no mode_____

19. 99, 102, 97, 110, 97, 93,
    98, 104, 108

    _____100.9; 99; 97_____

Place the decimal point in the product.

20. 27.95 × 4.3 = 120185

    _____120.185_____

21. 7.16 × 1.82 = 130312

    _____13.0312_____

22. 2.709 × 0.356 = 964404

    _____0.964404_____

Practice PW87

Name _____ LESSON 21.1

Percent

Write the percent that is shaded.

1.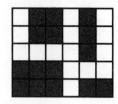
50%

2.
75%

3.
25%

4.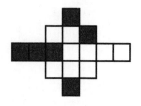
40%

5.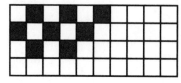
20%

6.
$66\frac{2}{3}$%

Write as a percent.

7. $\frac{87}{100}$ 87%
8. $\frac{6}{25}$ 24%
9. $\frac{9}{10}$ 90%
10. $\frac{120}{100}$ 120%
11. $\frac{13}{20}$ 65%
12. $\frac{1}{10}$ 10%
13. $\frac{7}{25}$ 28%
14. $\frac{85}{50}$ 170%

Compare. Write >, <, or =.

15. 2.3% __<__ 23%
16. 10% __>__ 7%
17. 5% __>__ 0.5%
18. 0.79% __<__ 7.9%
19. 125% __>__ 12.5%
20. 8.00% __=__ 8%

Mixed Review

Write the ratio in fraction form. Then find the unit rate.

21. 120 swimmers for 6 lifeguards $\frac{120}{6}$; 20 swimmers per lifeguard

22. 385 miles in 7 hours $\frac{385}{7}$; 55 mi per hr

23. $1.92 for 8 oz $\frac{1.92}{8}$; $0.24 per oz

24. $5.40 for a dozen muffins $\frac{5.40}{12}$; $0.45 per muffin

Write the prime factorization in exponent form.

25. 63 $3^2 \times 7$
26. 144 $3^2 \times 2^4$
27. 230 $2 \times 5 \times 23$

PW88 Practice

Name _____

LESSON 21.2

Percents, Decimals, and Fractions

Write as a percent.

1. 0.7 __70%__
2. 0.18 __18%__
3. 0.84 __84%__
4. 0.41 __41%__

5. $\frac{3}{5}$ __60%__
6. $\frac{17}{100}$ __17%__
7. $\frac{5}{8}$ __62.5%__
8. $\frac{8}{25}$ __32%__

Write each percent as a fraction or mixed number in simplest form.

9. 75% __$\frac{3}{4}$__
10. 30% __$\frac{3}{10}$__
11. 55% __$\frac{11}{20}$__
12. 240% __$2\frac{2}{5}$__

13. 6% __$\frac{3}{50}$__
14. 56% __$\frac{14}{25}$__
15. 105% __$1\frac{1}{20}$__
16. $12\frac{1}{2}$% __$\frac{1}{8}$__

Write each percent as a decimal.

17. 37% __0.37__
18. 9% __0.09__
19. 0.05% __0.0005__
20. 321% __3.21__

Compare. Write >, <, or =.

21. $\frac{1}{8}$ __>__ 8%
22. 23% __<__ 2.3
23. 30% __<__ $\frac{1}{3}$

Mixed Review

Find the quotient. Write the answer in simplest form.

24. $5 \div \frac{3}{8}$ __$13\frac{1}{3}$__
25. $6\frac{1}{3} \div \frac{2}{5}$ __$15\frac{5}{6}$__
26. $1\frac{1}{2} \div 3\frac{1}{3}$ __$\frac{9}{20}$__
27. $2\frac{3}{5} \div \frac{1}{8}$ __$20\frac{4}{5}$__

For 28–30, use the figure at the right.

28. Name two angles adjacent to ∠JOK.

    __∠POJ, ∠KOL__

29. Name an angle vertical to ∠KOL.

    __∠NOP__

30. Name two angles supplementary to ∠MON.

    __∠KOM, ∠NOJ__

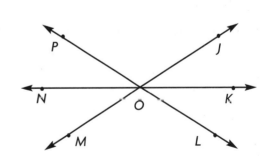

Practice   PW89

Name _____

LESSON 21.3

Estimate and Find Percent of a Number

Use a fraction in simplest form to find the percent of the number.

1. 10% of 8
   $\frac{4}{5}$

2. 25% of 60
   15

3. 50% of 50
   25

4. 70% of 90
   63

5. 80% of 70
   56

Use a decimal to find the percent of the number.

6. 15% of 8
   1.2

7. 35% of 45
   15.75

8. 55% of 92
   50.6

9. 82% of 70
   57.4

10. 93% of 24
    22.32

Use the method of your choice to find the percent of the number.

11. 52% of 40
    20.8

12. 96% of 84
    80.64

13. 81% of 34
    27.54

14. 12% of 300
    36

15. 67% of 200
    134

16. 4.5% of 90
    4.05

17. 110% of 30
    33

18. 140% of 100
    140

19. 200% of 250
    500

20. 400% of 80
    320

Estimate a 15% tip for each amount. Possible answers are given.

21. $12.00
    $1.80

22. $5.50
    $0.85

23. $23.75
    $3.60

24. $39.50
    $6.00

25. $94.80
    $15.00

Each proportion shows n as a percent of a number.
What is the percent? What is n?

26. $\frac{20}{100} = \frac{n}{40}$
    20%; 8

27. $\frac{5}{100} = \frac{n}{30}$
    5%; 1.5

28. $\frac{n}{150} = \frac{12}{100}$
    12%; 18

Mixed Review

Solve and check.

29. $5n = 45$
    $n = 9$

30. $\frac{m}{3} = 12$
    $m = 36$

31. $99 = {}^-9k$
    $k = {}^-11$

32. $21.4 = \frac{a}{6.3}$
    $a = 134.82$

Find the sum or difference. Write the answer in simplest form.

33. $\frac{5}{8} - \frac{1}{4}$
    $\frac{3}{8}$

34. $\frac{2}{5} + \frac{1}{3}$
    $\frac{11}{15}$

35. $\frac{5}{7} - \frac{1}{3}$
    $\frac{8}{21}$

36. $\frac{3}{4} + \frac{1}{10}$
    $\frac{17}{20}$

PW90 Practice

Name _____

LESSON 21.5

Discount and Sales Tax

Find the sale price.

1. regular price: $18.50
   [Discount 20%]
   $14.80

2. regular price: $35.00
   [25% off]
   $26.25

3. regular price: $45.50
   [SAVE 50%]
   $22.75

4. regular price: $23.60
   [SALE 80% off]
   $4.72

5. regular price $79.50
   discount rate: 15%
   $67.57

6. regular price $153.99
   discount rate: 10%
   $138.59

7. regular price $750.00
   discount rate: 18%
   $615.00

Find the regular price.

8. sale price $47.60
   discount rate: 30%
   $68.00

9. sale price $24.70
   discount rate: 5%
   $26.00

10. sale price $239.20
    discount rate: 20%
    $299.00

Find the sale tax for the given price. Round to the nearest cent.

11. $30.00
    tax: 8%
    $2.40

12. $15.80
    tax: 11%
    $1.74

13. $654.00
    tax: 7.5%
    $49.05

14. $1,842.00
    tax: 4%
    $73.68

Find the total cost of the purchase. Round to the nearest cent.

15. price: $79.50
    tax: 8%
    $85.86

16. price: $129.95
    tax: 6%
    $137.75

17. price: $405.00
    tax: 9%
    $441.45

18. price: $3,385.00
    tax: 5.5%
    $3,571.18

Mixed Review

Find the quotient.

19. $36 \div ^-4$ _____   20. $^-39 \div 3$ _____   21. $^-60 \div ^-15$ _____   22. $81 \div ^-9$ _____

Compare the fractions. Write >, <, or =.

23. $\frac{2}{3}$ ◯ $\frac{6}{8}$ 24. $\frac{10}{12}$ ◯ $\frac{5}{8}$ 25. $\frac{2}{7}$ ◯ $\frac{3}{5}$ 26. $\frac{8}{9}$ ◯ $\frac{13}{14}$

Practice   PW91

Name _____

LESSON 21.6

Simple Interest

Find the simple interest.

1. principal: $8,000
   rate: 5%
   time: 3 years

   _____ $1,200

2. principal: $1,500
   rate: 7.2%
   time: 10 years

   _____ $1,080

3. principal: $22,500
   rate: 4.8%
   time: 13 years

   _____ $14,040

Find the simple interest.

	Principal	Yearly Rate	Interest for 1 Year	Interest for 2 Years
4.	$80	3%	$2.40	$4.80
5.	$150	4.5%	$6.75	$13.50
6.	$340	6%	$20.40	$40.80
7.	$600	5.2%	$31.20	$62.40
8.	$1,400	7.9%	$110.60	$221.20
9.	$5,500	9%	$495.00	$990.00
10.	$7,500	8.5%	$637.50	$1,275.00
11.	$10,000	9.6%	$960.00	$1,920.00
12.	$11,350	9.8%	$1,112.30	$2,224.60
13.	$12,975	9.5%	$1,232.63	$2,465.25

Mixed Review

Convert the temperature to degrees Fahrenheit. Write the answer as a decimal.

14. 50°C   _____ 122°F
15. 10°C   _____ 50°F
16. 93°C   _____ 199.4°F
17. 23°C   _____ 73.4°F
18. 35°C   _____ 95°F

Find the sum.

19. $^+9 + {}^-6$ _____ 3
20. $^-4 + {}^-7$ _____ 11
21. $^-12 + {}^+8$ _____ 4
22. $^-20 + {}^+14$ _____ 6
23. $^+3 + {}^-18$ _____ 15
24. $^-6 + {}^-2$ _____ 8
25. $^-36 + {}^+36$ _____ 0
26. $^+30 + {}^-15$ _____ 15
27. $^-5 + {}^-8$ _____ 13

PW92 Practice

Name _____

LESSON 22.1

Theoretical Probability

Use the spinner at the right to find each probability. Write each answer as a fraction, a decimal, and a percent.

1. P(M) $\frac{1}{4}$, 0.25, 25%
2. P(H) $\frac{1}{8}$, 0.125, 12.5%
3. P(J) $\frac{0}{8}$, 0, 0%
4. P(T) $\frac{3}{8}$, 0.375, 37.5%
5. P(A) $\frac{1}{4}$, 0.25, 25%
6. P(M or A) $\frac{1}{2}$, 0.5, 50%
7. P(T or H) $\frac{1}{2}$, 0.5, 50%
8. P(M, A, or T) $\frac{7}{8}$, 0.875, 87.5%

A bag contains 5 blue, 3 red, and 2 green pencils. You choose one pencil without looking. Find each probability. Tell how likely the event is to occur. Write *impossible, unlikely, likely, very likely,* or *certain*.

9. P(pink) $\frac{0}{10}$
10. P(blue) $\frac{1}{2}$
11. P(green) $\frac{1}{5}$
12. P(blue or red) $\frac{4}{5}$

impossible   _likely_   _unlikely_   _very likely_

Cards numbered 2, 2, 2, 3, 4, 4, 5, and 5 are placed in a box. You choose one card without looking. Compare the probabilities. Write <, >, or = in each ◯.

13. P(2) > P(4)
14. P(4) = P(5)
15. P(3 or 5) < P(2, 3, or 5)

For 16–18, use the figure at the right. Find each probability.

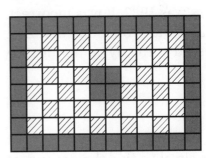

16. P(shaded square) $\frac{5}{12}$

17. P(striped or white square) $\frac{7}{12}$

18. P(shaded or striped square) $\frac{17}{24}$

Mixed Review

Evaluate the expression for $x = {}^-3, {}^-1,$ and 2.

19. $^-3x + 5$ 20. $x^2 - 4x$ 21. $7(2x + 1)$ 22. $x^2(6 - x)$

  14, 8, $^-1$ 21, 5, $^-4$ $^-35, ^-7, 35$ 81, 7, 16

Write the fraction as a percent.

23. $\frac{3}{4}$ 75%
24. $\frac{3}{10}$ 30%
25. $\frac{2}{25}$ 8%
26. $\frac{6}{5}$ 120%

Practice **PW93**

Name _____

LESSON 22.2

Problem Solving Skill: Too Much or Too Little Information

Write if each problem has *too much, too little,* or *the right amount* of information. Then solve the problem if possible, or describe the information needed to solve it.

1. It costs $1 to buy a drink from a machine. The machine has water, 3 types of juice, and 5 different sodas. If Caryn pushes one of the buttons without looking, what is the probability that she will get one of the juices?

   too much; $\frac{1}{3}$

2. Manny is in line at the Multiplex Theater. Of all the movies playing, there are 3 that Manny wants to see. If he buys a ticket without asking for a particular movie, what is the probability that he will get a ticket for a movie he wants to see?

   too little; need number of movies playing

3. Mr. Irving is playing a game at a charity carnival. He pays $15 for a chance to play. To find out what he has won, he reaches into a bag containing a $1 bill, a $5 bill, a $10 bill, a $20 bill, and a $50 bill. What is the probability that Mr. Irving will win more than the game cost?

   the right amount; $\frac{2}{5}$

4. Jessie ordered several books from an on-line store. When they arrived, she opened the carton, examined both science fiction books and the other novels. If she then randomly chose a book to read, what is the probability she chose one of the science fiction books?

   too little; need number of other books

5. Leah was trying to guess the year Ali was born. She knew it was anywhere from 1980 through 1985. Her first guess was 1982. It was incorrect. What is the probability that Leah guessed correctly on her next try?

   the right amount; $\frac{1}{5}$

6. Albert paid $8.95 for an almanac. He found out that in his city it rains an average of 75 days each year and snows an average of 15 days each year. What is the ratio of rainy days to snowy days?

   too much; 5 to 1

Mixed Review

Use a decimal to find the percent of the number.

7. 20% of 15

   3

8. 45% of 50

   22.5

9. 90% of 70

   63

10. 65% of 30

    19.5

Find the difference

11. $^+3 - {}^-7$

    $^+10$

12. $^+8 - {}^+15$

    $^-7$

13. $^-17 - {}^+5$

    $^-22$

14. $^-14 - {}^-9$

    $^-5$

PW94 Practice

Name _____

LESSON 22.4

Experimental Probability

Adam tossed a coin 50 times. For Exercises 1–2, use the table at the right to find the experimental probability.

Coin Toss	Heads	Tails
	22	28

1. P(Heads) __$\frac{11}{25}$__
2. P(Tails) __$\frac{14}{25}$__

3. What is the theoretical probability of getting heads? __$\frac{1}{2}$__

Sarah rolled a number cube numbered 1 to 6. The table below shows the results of rolling the cube 50 times. Use the results in the table to find the experimental probability.

Number	1	2	3	4	5	6
Times rolled	6	11	5	10	16	2

4. P(3) __$\frac{1}{10}$__
5. P(4 or 5) __$\frac{13}{25}$__
6. P(1 or 2) __$\frac{17}{50}$__

7. P(5) __$\frac{8}{25}$__
8. P(1) __$\frac{3}{25}$__
9. P(6) __$\frac{1}{25}$__

10. P(1 or 3) __$\frac{11}{50}$__
11. P(3 or 6) __$\frac{7}{50}$__
12. P(not 4) __$\frac{4}{5}$__

13. What is the theoretical probability for each number? __$\frac{1}{6}$__

Mixed Review

Multiply. Write the answer in simplest form.

14. $\frac{3}{4} \times \frac{2}{3}$ __$\frac{1}{2}$__
15. $\frac{1}{2} \times \frac{5}{6}$ __$\frac{5}{12}$__
16. $\frac{3}{8} \times \frac{4}{9}$ __$\frac{1}{6}$__
17. $\frac{5}{12} \times \frac{3}{10}$ __$\frac{1}{8}$__

Find the sum or difference. Write the answer in simplest form.

18. $1\frac{1}{2} + 3\frac{3}{8}$ __$4\frac{7}{8}$__
19. $5\frac{7}{8} - 2\frac{1}{4}$ __$3\frac{5}{8}$__
20. $\frac{7}{9} + 3\frac{2}{3}$ __$4\frac{4}{9}$__

21. $4\frac{2}{5} - 1\frac{3}{10}$ __$3\frac{1}{10}$__
22. $6\frac{1}{6} + 7\frac{3}{4}$ __$13\frac{11}{12}$__
23. $8\frac{5}{12} - 3\frac{1}{3}$ __$5\frac{1}{12}$__

Practice PW95

Name _____

LESSON 23.1

Problem Solving Strategy: Make an Organized List

Solve the problem by making an organized list.

1. Mr. Perez is planning a trip. He can leave on Monday, Wednesday, or Friday, at 8:00 A.M., 10:30 A.M., 2:00 P.M., or 4:30 P.M. How many choices does Mr. Perez have?

   _____ 12 choices _____

2. Len is going on vacation. He has 1 jacket, 2 sweaters, and 4 shirts. How many different outfits can Len make if each outfit consists of a jacket, sweater, and shirt?

   _____ 8 outfits _____

3. The 14 members of the bicycle team want to put 2-digit numbers on the backs of their jerseys. They decided to use only the digits 2, 4, 6, 8. Can each team member have a different number? How many possible 2-digit numbers are there?

   _____ yes; 16 2-digit numbers _____

4. Twelve members of the science club are planning their next field trip. They can take the trip in May or June. They can visit a science museum, bird sanctuary, zoo, or planetarium. How many different field trips involving 1 place and 1 month are possible?

   _____ 8 field trips _____

5. Ben is planning a hike to Eagle Mountain, Crystal Lake, and Cedar Falls. He needs to decide in which order to visit them. How many choices does he have?

   _____ 6 choices _____

6. Nina found jackets in blue, green, and red. She found scarves in yellow, beige, and navy. She wants to buy a jacket and scarf. How many different outfits can she choose from?

   _____ 9 outfits _____

Mixed Review

A number cube is labeled 2, 3, 5, 8, 9, 9. Find each probability.

7. P(5) __$\frac{1}{6}$__   8. P(9) __$\frac{1}{3}$__   9. P(even) __$\frac{1}{3}$__   10. P(not 9) __$\frac{2}{3}$__

Write a proportion. Then find the unknown length. The figures are similar.

11. [rectangles: n, 5.4 yd; 3.8 yd, 1.8 yd]

    $n = 11.4$ yd

12.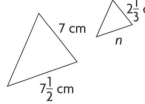

    $n = 2\frac{1}{2}$ cm

13.

    $n = 77$ m

PW96 Practice

Name _____

LESSON 23.2

Compound Events

Draw a tree diagram or make a table to find the number of possible outcomes for each situation. Check students' diagrams or tables.

1. spinning a pointer on a spinner labeled 1 to 4 and tossing a coin

   _____8 outcomes_____

2. a choice of 3 cards, 2 envelopes, and 2 stickers

   _____12 outcomes_____

3. a choice of either a red, blue, or green shirt and a black, gray, or brown jacket

   _____9 outcomes_____

4. a choice of 4 sandwiches, 2 drinks, and 2 desserts

   _____16 outcomes_____

Use the Fundamental Counting Principle to find the number of possible outcomes for each situation.

5. a choice of 3 juices, 2 muffins, and 3 sandwiches

   _____18 outcomes_____

6. a choice of 3 beverages, 2 snacks, and 4 sandwiches

   _____24 outcomes_____

7. tossing a coin and rolling 2 number cubes labeled 1 to 6.

   _____72 outcomes_____

8. a choice of 4 shirts, 4 ties, 3 trousers, and 3 belts

   _____144 outcomes_____

9. A Chinese restaurant offers 25 main dishes, 3 kinds of rice, and 3 different beverages. If the restaurant is open every day of the year, is it possible to eat a different meal there every day for a year? Explain.

   no; $25 \times 3 \times 3 = 225$ and $225 < 365$

10. Mr. Samson is building a house. It will be in either New Jersey or New York. He wants either a one-story or a two-story house with either 3 or 4 bedrooms. How many choices does he have?

    _____8 choices_____

Mixed Review

Write each percent as a decimal.

11. 28% ____0.28____
12. 5% ____0.05____
13. 163% ____1.63____
14. 91% ____0.91____

Find the measure of the third angle of the triangle and classify the triangle.

15. $\angle 1 = 36°$; $\angle 2 = 18°$
    $\angle 3 = $ ▮°

    _____126°; obtuse_____

16. $\angle 1 = 53°$; $\angle 2 = 37°$
    $\angle 3 = $ ▮°

    _____90°; right_____

17. $\angle 1 = 68°$; $\angle 2 = 59°$
    $\angle 3 = $ ▮°

    _____53°; acute_____

Practice PW97

Name _____

LESSON 23.3

Independent and Dependent Events

Write *independent* or *dependent* to describe the events.

1. roll two number cubes two times

   _____independent_____

2. select a lettered tile from a box, do not replace it, select another tile

   _____dependent_____

3. select a coin from a jar, do not replace it, select another coin

   _____dependent_____

4. select a marble from a bag, replace it, select another marble

   _____independent_____

Without looking, you take a card out of a jar and replace it before selecting again. Find the probability of each event. Then find the probability assuming the card is not replaced after each selection.

5. P(5, 6)

   $\frac{2}{25}, \frac{1}{10}$

6. P(6, 8)

   $\frac{1}{25}, \frac{1}{20}$

7. P(5, 7 or 8)

   $\frac{4}{25}, \frac{1}{5}$

8. P(6, not 5)

   $\frac{3}{25}, \frac{1}{10}$

9. P(6, 7 or 8)

   $\frac{2}{25}, \frac{1}{10}$

10. P(5, even)

    $\frac{4}{25}, \frac{1}{5}$

11. P(7, 7)

    $\frac{1}{25}, 0$

12. P(7, 6, 5)

    $\frac{2}{125}, \frac{1}{30}$

13. P(6, odd)

    $\frac{3}{25}, \frac{3}{20}$

14. P(5, 8, 5)

    $\frac{4}{125}, \frac{1}{30}$

15. P(5, 6, 7)

    $\frac{2}{125}, \frac{1}{30}$

16. P(5, 5)

    $\frac{4}{25}, \frac{1}{10}$

Mixed Review

Use the formula $d = rt$ to complete.

17. $d = 858$ cm
    $r = 55$ cm per second
    $t = \blacksquare$ sec

    _____15.6 sec_____

18. $r = 48$ mi per hr
    $t = 6.5$ hr
    $d = \blacksquare$ mi

    _____312 mi_____

19. $d = 423$ m
    $t = 18$ min
    $r = \blacksquare$ m per min

    _____23.5 m per min_____

Write the fraction in simplest form.

20. $\frac{25}{150}$ $\frac{1}{6}$

21. $\frac{42}{210}$ $\frac{1}{5}$

22. $\frac{27}{72}$ $\frac{3}{8}$

23. $\frac{39}{52}$ $\frac{3}{4}$

Name _____

LESSON 23.4

Make Predictions

The results of a survey of 600 randomly selected teenagers in California indicate that 150 of them use their computers at least 2 hr a day.

1. What is the probability that a randomly selected teenager in California uses the computer at least 2 hr a day?

   $\frac{1}{4}$, 0.25, or 25%

2. Out of 5,500 California teenagers, predict about how many would indicate that they use their computers at least 2 hr a day.

   about 1,375 teenagers

3. In a sample of 800 bicycles, the quality control department found that 32 of them were defective. If the company manufactures 8,000 bicycles, about how many of them will be defective?

   about 320 bicycles

4. In a sample of 700 phones, the quality control department found that 21 of them were defective. If the company manufactures 14,000 phones, about how many of them will be defective?

   about 420 phones

Use Data For 5–6, use the table. The table shows the favorite sports indicated by a random sample of 150 sixth graders from Glenville Middle School.

5. If there are 360 sixth graders at Glenville Middle School, about how many will prefer baseball? soccer?

   about 108 sixth graders, about 60 sixth graders

Favorite Sports

Sport	Number of Students
baseball	45
basketball	20
soccer	25
hockey	10
track	15
football	35

6. If there are 450 sixth graders at Glenville Middle School, about how many will prefer a sport other than baseball?

   about 315 sixth graders

Mixed Review

Find the difference.

7. $^+8 - {^+9}$ ___−1___   8. $^-5 - {^+3}$ ___−8___   9. $^+12 - {^-4}$ ___+16___   10. $^-6 - {^-9}$ ___+3___

Find the LCM of each pair of numbers.

11. 5, 13 ___65___   12. 12, 18 ___36___   13. 9, 15 ___45___   14. 8, 22 ___88___

Practice PW99

Name _____

LESSON 24.1

Algebra: Customary Measurements

Use a proportion to convert to the given unit.

1. 80 fl oz = __10__ c
2. 18 pt = __9__ qt
3. 510 ft = __170__ yd

4. 720 in. = __20__ yd
5. 5 months ≈ __20__ weeks
6. 6 c = __48__ fl oz

7. 3 gal = __12__ qt
8. 6 T = __12,000__ lb
9. 32 fl oz = __4__ c

10. 5 mi = __26,400__ ft
11. 44 qt = __11__ gal
12. 10 pt = __20__ c

13. 157 ft = __52__ yd __1__ ft
14. 220 in. = __18__ ft __4__ in.
15. $5\frac{2}{3}$ yd = __17__ ft

16. $5\frac{1}{2}$ T = __11,000__ lb
17. $6\frac{1}{4}$ ft = __75__ in.
18. $7\frac{1}{2}$ ft = __$2\frac{1}{2}$__ yd

19. 325 ft = __108__ yd __1__ ft
20. $3\frac{3}{4}$ yd = __$11\frac{1}{4}$__ ft
21. 15 gal = __60__ qt

Compare. Write <, >, or = for each ●.

22. 8,500 lb ● 4 T
__>__

23. 25 yd ● 75 ft
__=__

24. 9 days ● 225 hrs
__<__

25. 16 c ● 4 qt
__=__

26. 5 gal ● 30 pt
__>__

27. 12 ft ● 120 in.
__>__

Mixed Review

Solve and check.

28. 6x = 84 __x = 14__
29. $\frac{w}{11}$ = 6 __w = 66__
30. 3.5 m = 14 __m = 4__

31. $\frac{c}{8}$ = 4.7 __c = 37.6__
32. 6.3 = $\frac{h}{20}$ __h = 126__
33. 9.9 = 1.8 r __r = 5.5__

Find the product.

34. ⁻11 × ⁻5 __55__
35. 18 × ⁻6 __⁻108__
36. ⁻7 × 8 __⁻56__
37. ⁻14 × ⁻21 __294__

PW100 Practice

LESSON 24.2

Name _____

Algebra: Metric Measurements

Complete the pattern.

1. 1 L = __100__ cL
2. 1,000 mg = __1__ g
3. 1 m = __0.001__ km

   0.1 L = __10__ cL
   100 mg = __0.1__ g
   10 m = __0.01__ km

   0.01 L = __1__ cL
   10 mg = __0.01__ g
   100 m = __0.1__ km

   1 mg = __0.001__ g
   1,000 m = __1__ km

Use a proportion to convert to the given unit.

4. 40 g = __0.04__ kg
5. 300 km = __300,000__ m
6. 9 kL = __9,000__ L
7. 6 kL = __60,000__ dL
8. 300 cm = __30__ dm
9. 50 dL = __500__ cL
10. 12 kL = __12,000__ L
11. 28 g = __28,000__ mg
12. 8 km = __8,000__ m
13. 2.2 g = __220__ cg
14. 7 dm = __0.7__ m
15. 5.5 cg = __0.55__ dg

Compare. Write , < >, or = for ●.

16. 600 mm ● 6 m   __<__
17. 80 km ● 80,000 m   __=__
18. 4,000 mL ● 4 L   __=__

19. 2.5 kg ● 25,000 mg   __>__
20. 50 kL ● 50,000 L   __=__
21. 14,500 mg ● 145 g   __<__

Mixed Review

Find the quotient.

22. ⁻85 ÷ ⁻5   __17__
23. ⁻48 ÷ 8   __6__
24. 162 ÷ ⁻9   __18__
25. ⁻132 ÷ ⁻12   __11__

Write the difference in simplest form.

26. $7\frac{1}{2} - 6\frac{1}{3}$ $1\frac{1}{6}$
27. $5\frac{5}{8} - 3\frac{3}{4}$ $1\frac{7}{8}$
28. $10\frac{1}{6} - 5\frac{7}{8}$ $4\frac{7}{24}$
29. $15\frac{1}{3} - 12\frac{4}{5}$ $2\frac{8}{15}$

Practice PW101

Name _____

LESSON 24.3

Algebra: Relate Customary and Metric

Use a proportion to convert to the given unit.

1. 4 ft ≈ __?__ cm

2. 16 yd ≈ __?__ m

3. 12 qt ≈ __?__ L

    121.92   14.56   11.4

4. 30 lb ≈ __?__ kg

5. 16 L ≈ __?__ qt

6. 64 cm ≈ __?__ ft

    13.5   16.84   2.1

7. 3 in. ≈ __?__ mm

8. 130 yd ≈ __?__ m

9. 120 L ≈ __?__ gal

    76.2   118.3   31.66

10. 52 m ≈ __?__ ft

11. 150 kg ≈ __?__ lb

12. 6 m ≈ __?__ ft

    170.56   333.33   19.68

13. 2 ft ≈ __?__ cm

14. 40 yd ≈ __?__ m

15. 3 oz ≈ __?__ g

    60.96   36.4   85.05

Compare. Write <, >, or = for each ●.

16. 7 ft ● 421 cm

17. 80 cm ● 4.5 ft

18. 8.5 lb ● 5 kg

    <   <   <

19. 44 mm ● 5 in.

20. 32.8 ft ● 9 m

21. 5 km ● 3.2 mi

    <   >   <

22. 22 gal ● 55 L

23. 8.2 qt ● 10.1 L

24. 1.5 mi ● 1.7 km

    >   <   >

Mixed Review

Solve and check.

25. $x + 7 = 19$

26. $y - 9 = 7$

27. $m + 19 = 41$

    $x = 12$ $y = 16$ $m = 22$

28. $r - 27 = 15$

29. $z + 4.7 = 11$

30. $w - 7.8 = 5.6$

    $r = 42$ $z = 6.3$ $w = 13.4$

Find the difference.

31. $^{-}5 - 7$ $^{-}12$

32. $8 - 10$ $^{-}2$

33. $6 - ^{-}9$ 15

PW102 Practice

Name _____

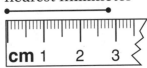

Appropriate Tools and Units

Measure the line segment to the given length.

1. nearest inch; nearest half inch

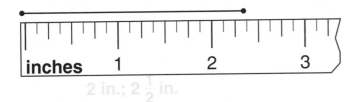

   2 in.; 2 1/2 in.

2. nearest centimeter; nearest millimeter

   3 cm; 28 mm

3. nearest half inch; nearest quarter inch

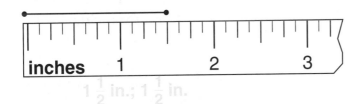

   1 1/2 in.; 1 1/2 in.

4. nearest centimeter; nearest millimeter

   5 cm; 53 mm

5. nearest inch; nearest half inch

   2 in.; 2 in.

6. nearest centimeter; nearest millimeter

   2 cm; 19 mm

Tell which measurement is more precise.

7. 9 lb or 142 oz   142 oz

8. 6 c or 50 fl oz   50 fl oz

9. 40 cm or 420 mm   420 mm

10. 9 mg or 0.8 cg   9 mg

11. 2 yd or 71 in.   71 in.

12. 1 L or 980 mL   980 mL

Name an appropriate customary or metric unit of measure for each item.

13. the amount of formula in a baby's bottle   fluid ounces or milliliters

14. the weight of a laptop computer   pounds or kilograms

15. the length of the eraser on a pencil   millimeters or parts of an inch

16. the weight of a box of tissues   ounces or grams

Mixed Review

Solve and check.

17. $5w = 30$ $w = 6$

18. $\frac{m}{4} = 5$ $m = 20$

19. $72 = 9h$ $h = 8$

20. $\frac{w}{3} = 17$ $w = 51$

Write the ratio in three ways.

21. five to nine   5 to 9; 5:9; $\frac{5}{9}$

22. ten to seven   10 to 7; 10:7; $\frac{10}{7}$

Practice PW103

Name _____

LESSON 24.5

Problem Solving Skill: Estimate or Find Exact Answer

Decide whether you need an estimate or an exact answer. Solve.

1. You and several friends are setting up tents on a camping trip. It takes 25 min to set up a tent. If you begin at 1:00 P.M., can you set up 7 tents by 3:00 P.M.?

   estimate; no

2. Your campsite is a rectangle 61 ft by 33 ft. You have 200 ft of rope. Do you have enough to run the rope around the entire perimeter of the campsite?

   estimate; yes

3. You brought 9 bags of snacks with you for the 2-day trip. Each bag cost $1.59. If you paid for the snacks with a $20 bill, how much change did you receive?

   exact; $5.69

4. On the second day of the trip, your group hikes for $3\frac{3}{4}$ hr. If you average 3.8 mi per hr, will you have reached your goal of 10 mi for the day?

   estimate; yes

5. The odometer on the van you rented for the trip read 5,398.2 mi when you left home. It read 5,702.1 mi when you arrived back home. How far did you drive?

   exact; 303.9 mi

6. Everyone agrees that they want to get at least 8 hr sleep per night. If you want to wake up at 6:45 A.M. each morning, what is the latest you can fall asleep each night?

   exact; 10:45 P.M.

7. The trip to the campground usually takes about $3\frac{1}{4}$ hr. If you leave home at 8:45 A.M. and make two 20-min stops, would you arrive by noon?

   estimate; no

8. On the last night of the camping trip, you have 1 gal of water left. After making 3 cans of soup that each required 16 fl oz of water, how many fluid ounces of water do you have left?

   exact; 80 fl oz

Mixed Review

Write the sum or difference in simplest form.

9. $\frac{1}{2} - \frac{1}{3}$
   $\frac{1}{6}$

10. $\frac{2}{5} - \frac{1}{4}$
    $\frac{3}{20}$

11. $\frac{2}{5} - \frac{1}{6}$
    $\frac{7}{30}$

12. $\frac{5}{8} - \frac{1}{4}$
    $\frac{3}{8}$

13. $\frac{7}{8} - \frac{3}{4}$
    $\frac{1}{8}$

14. $\frac{7}{10} + \frac{1}{5}$
    $\frac{9}{10}$

15. $\frac{1}{6} + \frac{1}{3}$
    $\frac{1}{2}$

16. $\frac{3}{5} + \frac{1}{3}$
    $\frac{14}{15}$

17. $\frac{1}{9} + \frac{1}{2}$
    $\frac{11}{18}$

18. $\frac{2}{9} + \frac{1}{3}$
    $\frac{5}{9}$

Find the sum.

19. $^-3 + {}^-7$
    _$^-10$_

20. $^-4 + 6$
    _2_

21. $5 + {}^-8$
    _$^-3$_

22. $4 + 7$
    _11_

PW104 Practice

Name _____

LESSON 25.2

Perimeter

Find the perimeter.

1.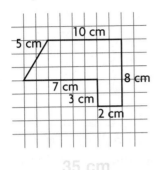

   35 cm

2.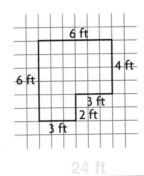

   24 ft

3.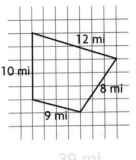

   39 mi

Find the unknown length.
Then find the perimeter.

4.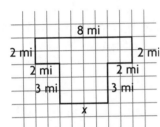

   $x = 4$ mi;

   26 mi

The perimeter is given.
Find the unknown length.

5.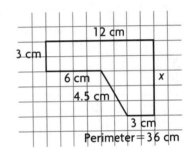

   $x = 7.5$ cm

Mixed Review

Use a proportion to change to the given unit.

6. 18 yd = __648__ in.

7. 22 qt = __$5\frac{1}{2}$__ gal

8. 435 min = __$7\frac{1}{4}$__ hr

9. 8.5 gal = __68__ pt

10. 348 in. = __29__ ft

11. $12\frac{1}{2}$ lb = __200__ oz

Write a numerical or algebraic expression for the word expression.

12. One hundred divided by the sum of k and m. __$100 \div (k + m)$, or $\frac{100}{k+m}$__

13. v less than two thousand forty-seven. __$2{,}047 - v$__

14. w multiplied by the product of a and b. __$w \times a \times b$, or $w(ab)$__

Practice PW105

Name _____

LESSON 25.3

Problem Solving Strategy: Draw a Diagram

Solve the problem by drawing a diagram.

A contractor built a house in the shape of a rectangle. The house is 64 ft long and 48 ft wide. There is a wall running across the width of the house. The wall divides the length of the house into two sections, one larger than the other. The distance from the wall to one end of the house is 3 times the distance from the wall to the other end.

1. Describe the shape of the larger section of the house. Give the dimensions of the figure.

   __square; each side is 48 ft__

2. Molding is going to be installed around the entire floor of the larger section of the house. How many feet of molding will be needed?

   __192 ft__

3. There are 3 doors leading into the house. Each door is 3 ft wide. What is the perimeter of the house if the doors are not included?

   __215 ft__

4. There are beams around the perimeter of the house every 16 in. If there is a beam in each corner, what is the total number of beams?

   __168 beams__

5. There are plans to add a garage to the side of the house. The length of the rectangular garage will be 3 ft greater than its width. If the perimeter of the garage will be 90 ft, find its length and width.

   __24 ft; 21 ft__

6. The short side of the garage will be attached to a short side of the house. What will be the perimeter of the house and garage when the garage is complete?

   __272 ft__

Mixed Review

Write the ratio as a percent.

7. $\frac{74}{100}$ __74%__

8. $\frac{6}{100}$ __6%__

9. $\frac{19}{100}$ __19%__

10. $\frac{40}{100}$ __40%__

11. $\frac{1}{100}$ __1%__

12. $\frac{100}{100}$ __100%__

13. $\frac{50}{100}$ __50%__

14. $\frac{98}{100}$ __98%__

Use a proportion to change to the given unit.

15. 2,400 mL = __2.4__ L

16. 5.8 kg = __5,800__ g

17. 150 cm = __1.5__ m

18. 0.7 g = __700__ mg

19. 1,300 m = __1.3__ km

20. 7,500 L = __7.5__ kL

Name _____

LESSON 25.4

Circumference

Find the circumference of the circle. Use 3.14 or $\frac{22}{7}$ for π. Round to the nearest whole number.

1.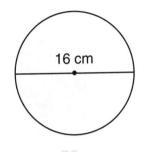

   50 cm

2.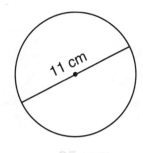

   35 mm

3.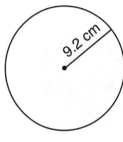

   58 cm

4.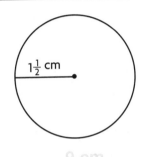

   9 cm

5.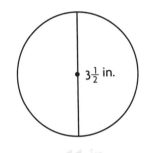

   11 in.

6.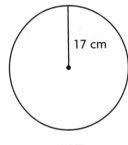

   107 cm

7.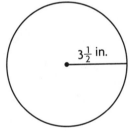

   22 in.

8.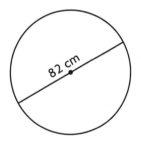

   257 cm; or 258 cm

9.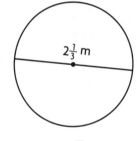

   7 m

Mixed Review

Use inverse operations to solve. Check your solution.

10. $3x = 12$

    $x = 4$

11. $40 = 8m$

    $m = 5$

12. $\frac{y}{3} = 6$

    $y = 18$

13. $\frac{h}{7} = 6$

    $h = 42$

Find the sale price.

14. regular price:
    $48.00
    20% off

    $38.40

15. regular price:
    $72.00
    30% off

    $50.40

16. regular price:
    $120.00
    60% off

    $48.00

17. regular price:
    $95.00
    25% off

    $71.25

Practice   **PW107**

Name _____

LESSON 26.1

Estimate and Find Area

Estimate the area of the figure. Each small square on the grid represents 1 in.²
Possible estimates are given.

1.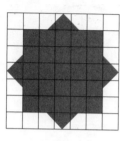

   about 28 in.²

2.

   about 19 in.²

3.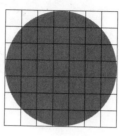

   about 38 in.²

4.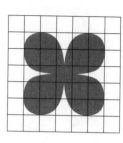

   about 20 in.²

Find the area.

5.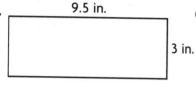

   28.5 in.²

6.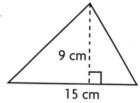

   67.5 cm²

7.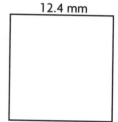

   153.76 mm²

8.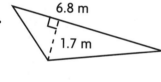

   5.78 m²

9.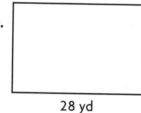

   602 yd²

10.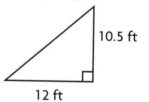

    63 ft²

Mixed Review

Find the circumference of the circle. Use 3.14 for π.

11. $d = 17$ cm   12. $d = 3.5$ in.   13. $r = 11$ mm   14. $r = 6.1$ ft

   53.38 cm   10.99 in.   69.08 mm   38.308 ft

Solve and check.

15. $x + 7 = 19$ 16. $30 = a + 13$ 17. $^-45 = 22.5 + n$ 18. $c + 2.3 = 9.1$

    x = 12   a = 17   n = ⁻67.5   c = 6.8

PW108 Practice

Name _____

LESSON 26.2

Algebra: Areas of Parallelograms and Trapezoids

Find the area of each figure.

1.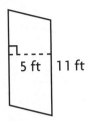

2.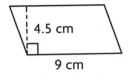

3.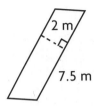

_____55 ft²_____   _____40.5 cm²_____   _____15 m²_____

4.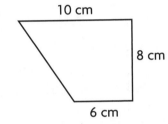

5.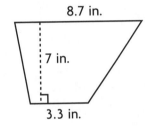

6.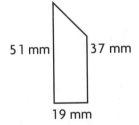

_____64 cm²_____   _____42 in²_____   _____836 mm²_____

7.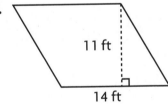

8.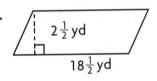

9.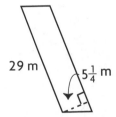

_____154 ft²_____   _____46.25 yd²_____   _____152.25 m²_____

Mixed Review

Tell which measurement is more precise.

10. 1 ft or 10 in.   11. 2 T or 2,010 lb   12. 3 qt or 1 gal   13. 2 kg or 2,020 g

_____10 in._____   _____2,010 lb_____   _____3 qt_____   _____2,020 g_____

Evaluate the expression for $x = {}^-2, 0,$ and 4.

14. $6 - \dfrac{x}{2}$ 15. $5x + 12$ 16. $^-2 - 3x$ 17. $(^-x + 2) \cdot 3$

_____7, 6, 4_____   _____2, 12, 32_____   _____4, ^-2, ^-14_____   _____12, 6, ^-6_____

Practice   PW109

Name _____   LESSON 26.4

Algebra: Areas of Circles

Find the area of each circle to the nearest whole number.
All areas are approximations.

1. 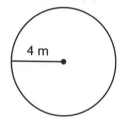   2. 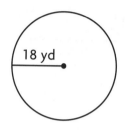   3.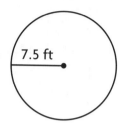

   _____50 m²_____   _____1,017 yd²_____   _____177 ft²_____

4. $r = 17$ yd __907 yd²__   5. $d = 38$ ft __1,134 ft²__   6. $r = 5.6$ m __98 m²__

7. $d = 10$ mm __79 mm²__   8. $r = 2.2$ mi __15 mi²__   9. $d = 54$ cm __2,289 cm²__

10. $r = 21$ ft __1,385 ft²__   11. $d = 1.8$ mi __3 mi²__   12. $r = 15.5$ in. __754 in.²__

13. $d = 30$ cm __707 cm²__   14. $r = 6.6$ yd __137 yd²__   15. $d = 16$ m __201 m²__

Find the area of the semicircle or quarter circle to the nearest whole number. Use 3.14 for π. All areas are approximations.

16.   17. 6.5 yd

   22 mm

   _____190 mm²_____   _____33 yd²_____

Mixed Review

A number cube is labeled 1 to 6. Find each probability.

18. P(3)   19. P(1 or 6)   20. P(8)   21. P(even)

    $\frac{1}{6}$ $\frac{1}{3}$ 0 $\frac{1}{2}$

Compare the numbers. Write <, >, or = for each ◯.

22. 0.01 < 0.11   23. 19.9 = 19.90   24. 0.411 > 0.401   25. 1.575 < 1.757

PW110 Practice

Name _____

LESSON 26.5

Algebra: Surface Areas of Prisms and Pyramids

Find the surface area.

1.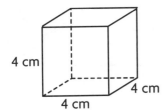

   __96 cm²__

2.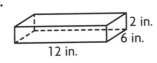

   __216 in.²__

3.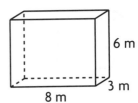

   __180 m²__

4.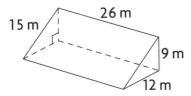

   __1,096 m²__

5.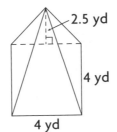

   __36 yd²__

6.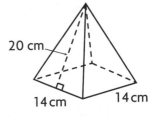

   __756 cm²__

7.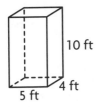

   __220 ft²__

8.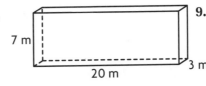

   __442 m²__

9.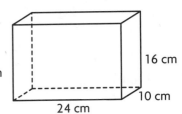

   __1,568 cm²__

Mixed Review

Evaluate the expression for $m = 6$ and $n = {}^-2$.

10. $m \div 3 - 4n$

    __10__

11. $(50 - m^2) \times 3 + n$

    __40__

12. $12n - 5 \times {}^-2m$

    __36__

Find the LCM of each pair of numbers.

13. 4, 18 __36__

14. 6, 32 __96__

15. 3, 11 __33__

Practice   PW111

Name _____

LESSON 27.1

Estimate and Find Volume

Find the volume.

1.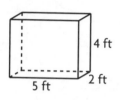

_____40 ft³_____

2.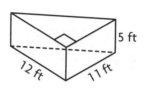

_____330 ft³_____

3.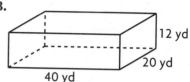

_____9,600 yd³_____

4.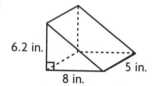

_____124 in.³_____

5.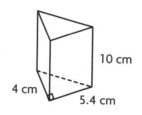

_____108 cm³_____

6.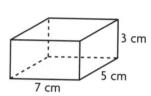

_____105 cm³_____

Find the unknown length.

7.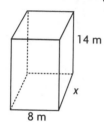
$V = 672$ m³

_____x = 6 m_____

8.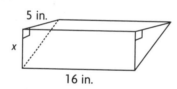
$V = 160$ in.³

_____x = 4 in._____

9.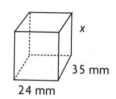
$V = 17,640$ mm³

_____x = 21 mm_____

Mixed Review

Find the circumference of the circle to the nearest whole number.
Use 3.14 or $\frac{22}{7}$ for π.

10. $r = 4$ in.

_____about 25 in._____

11. $d = 6.3$ cm

_____about 20 cm_____

12. $r = 12\frac{1}{2}$ m

_____about 79 m_____

13. $d = 9\frac{1}{3}$ yd

_____about 29 yd_____

14. $r = 110$ mm

_____about 691 mm_____

15. $d = 15.7$ ft

_____about 49 ft_____

Find the unknown dimension.

16. scale: 1 cm:12 m
    drawing length: 18 cm
    actual length: __216 m__

17. scale: 3 cm:2 mm
    drawing length: __16.5 cm__
    actual length: 11 mm

PW112 Practice

Name _____

LESSON 27.2

Problem Solving Strategy: Make a Model

Find the volume. Then double the dimensions. Find the new volume.

1.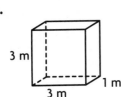
2.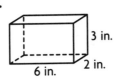
3.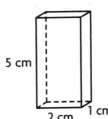
4.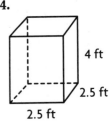

   9 m³; 72 m³   36 in.³; 288 in.³   10 cm³; 80 cm³   25 ft³; 200 ft³

Find the volume of each prism. Then halve the underlined dimension and find the new volume.

	Length	Width	Height	Volume	New Volume
5.	5 m	4 m	2 m	40 m³	20 m³
6.	12 ft	8 ft	10 ft	960 ft³	480 ft³
7.	24 cm	3 cm	6 cm	432 cm³	216 cm³
8.	9 in.	6 in.	10 in.	540 in.³	270 in.³

Mixed Review

A number cube is numbered 1 through 6. Find each probability.

9. P(5) $\frac{1}{6}$ 10. P(not 2) $\frac{5}{6}$ 11. P(even) $\frac{1}{2}$ 12. P(1 or 3) $\frac{1}{3}$

Solve and check.

13. $n + 8 = 57$
    $n = 49$

14. $30 = x + 82$
    $x = {}^-52$

15. $k + 2\frac{1}{3} = 11$
    $k = 8\frac{2}{3}$

16. $22 = 7.34 + b$
    $b = 14.66$

17. $17\frac{1}{2} + m = 23\frac{1}{4}$
    $m = 5\frac{3}{4}$

18. $^-44 = 37 + a$
    $a = {}^-81$

Practice   PW113

Name _____

LESSON 27.3

Algebra: Volumes of Pyramids

Find the volume.

1.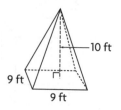

   _____270 ft³_____

2.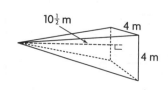

   _____56 m³_____

3.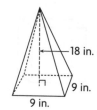

   _____486 in.³_____

4.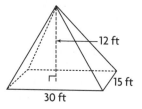

   _____1,800 ft³_____

5.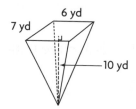

   _____140 yd³_____

6.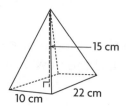

   _____1,100 cm³_____

7. rectangular pyramid: $l = 36$ in., $w = 50$ in., $h = 60$ in.

   _____36,000 in.³_____

8. square pyramid: $l = 14$ yd, $w = 14$ yd, $h = 25$ yd

   _____$1,633\frac{1}{3}$ yd³_____

9. rectangular pyramid: $l = 6$ cm, $w = 5$ cm, $h = 6$ cm

   _____60 cm³_____

10. square pyramid: $l = 10$ m, $w = 10$ m, $h = 15$ m

    _____500 m³_____

Mixed Review

Find the area of each circle to the nearest whole number. Use 3.14 for π.

11. $r = 4$ yd

    __about 50 yd²__

12. $d = 12$ ft

    __about 113 ft²__

13. $r = 5.5$ m

    __about 95 m²__

14. $d = 7.2$ cm

    __about 41 cm²__

Use a proportion to change to the given unit.

15. 4 qt = __16__ c

16. 26 in. = __$2\frac{1}{6}$__ ft

17. 68 oz = __$4\frac{1}{4}$__ lb

PW114 Practice

Name _____

LESSON 27.5

Volumes of Cylinders

Find the volume. Round to the nearest whole number.

1.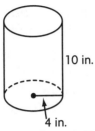

   about 502 in.³

2.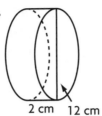

   about 226 cm³

3.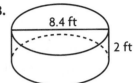

   about 111 ft³

4.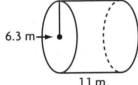

   about 1,371 m³

5.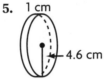

   about 66 cm³

6.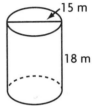

   about 3,179 m³

Find the volume of the inside cylinder to the nearest whole number.

7.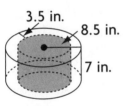

   550 in.³

8.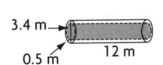

   54 m³

9.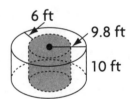

   453 ft³

Mixed Review

There are 6 blue socks, 5 red socks, and 10 black socks in a drawer. Without looking, you pull out 2 socks. Find the probabilities.

10. P(red, black) $\frac{5}{42}$

11. P(blue, blue) $\frac{1}{14}$

12. P(black, blue) $\frac{1}{7}$

Find the sum.

13. $^-7 + {}^-8$ 14. $^-3 + 14$ 15. $^-9 + {}^-27$ 16. $^-15 + 3$ 17. $22 + {}^-18$

    $^-1$ 11 $^-36$ $^-12$ 4

Practice **PW115**

Name _____

LESSON 28.1

Problem Solving Strategy: Find a Pattern

Solve the problems by finding a pattern.

1. Laura read a novel she found in the school library. She read 15 pages the first day. Then each day she read 6 more pages than the day before. How many pages did she read on the eighth day?

   _____57 pages_____

2. When Jeff played his new computer game for the first time, he scored 10,000 points. Each time he played, he increased his score by 15,000 points. How many games did Jeff have to play to reach a score of 100,000?

   _____7 games_____

3. For her pet store's grand opening, Mrs. Santos gave 7 prizes. The seventh-prize winner received a $1 gift certificate, the sixth-prize winner a $2 certificate, the fifth-prize winner a $4 certificate, the fourth-prize winner an $8 certificate. What was the value of the first-prize certificate?

   _____$64_____

4. Kevin is laying tile in his kitchen. The area of the kitchen is 96 ft². Since this is his first tile job, he is working at it slowly. He tiled 7 ft² the first day, 14 ft² the second day, and 21 ft² the third day. If this pattern continues, how many days will it take Kevin to tile the entire floor?

   _____5 days_____

5. The school band is practicing for a competition to be held in 8 weeks. The band practices 1 hr a day for the first week. It practices $1\frac{1}{4}$ hr a day the second week, $1\frac{1}{2}$ hr a day the third week, and $1\frac{3}{4}$ hr a day the fourth week. If this pattern continues, how many hours a day will the band practice during the eighth week?

   _____$2\frac{3}{4}$ hr_____

6. A team of synchronized swimmers makes patterns in the water by hooking their arms and legs together. One swimmer begins the formation. After 5 sec, two swimmers join. At 10 sec, two more join. At 15 sec, another two join the group. If this pattern continues, how many swimmers will be in the group after 30 sec?

   _____13 swimmers_____

Mixed Review

Find the volume of the cylinder. Round to the nearest whole number.

7. diameter = 13 in.
   height = 5 in.

   _____663 in.³_____

8. radius = 7.5 cm
   height = 24 cm

   _____4,239 cm³_____

9. diameter = 30 m
   height = 73 m

   _____51,575 m³_____

Find the value of the expression.

10. $24 + \sqrt{64} - 6^2 - \sqrt{81}$

    _____⁻13_____

11. $\sqrt{144} + 7^2 - \sqrt{36} - 5^2$

    _____30_____

PW116 Practice

Patterns in Sequences

Write a rule for each sequence.

1. 17, 22, 28, 35, . . .
   Add increasing integers starting with 5.

2. 81, 69, 57, 45, . . .
   Subtract 12 from each term.

3. 1, 5, 25, 125, . . .
   Multiply each term by 5.

4. 117, 116, 113, 108, . . .
   Subtract consecutive odd integers.

5. 700, 70, 7, 0.7, . . .
   Divide each term by 10.

6. 1,000, 500, 250, 125, . . .
   Divide each term by 2.

7. 77, 79, 83, 85, 89, . . .
   Add 2, 4, 2, 4, 2, 4, . . .

8. 19, 16.5, 14, 11.5, . . .
   Subtract 2.5 from each term.

9. 64, 55, 47, 40, . . .
   Subtract decreasing integers starting with 9.

Find the next three possible terms in each sequence.

10. 17, 34, 68, 136, . . .
    272; 544; 1,088

11. 325, 320, 310, 295, . . .
    275, 250, 220

12. 14.6, 14.5, 14.3, 14.0, . . .
    13.6, 13.1, 12.5

13. 3, 9, 27, 81, . . .
    243; 729; 2,187

14. 535, 529, 522, 514, . . .
    505, 495, 484

15. 33, 45, 57, 69, . . .
    81, 93, 105

16. 1,458, 486, 162, 54, . . .
    18, 6, 2

17. 390, 401, 414, 429, . . .
    446, 465, 486

18. 7, ⁻14, 28, ⁻56, . . .
    112, ⁻224, 448

Mixed Review

Find the surface area.

19. rectangular prism
    l = 18 cm
    w = 14 cm
    h = 8 cm
    1,016 cm²

20. cube
    edge = 5.5 in.
    181.5 in.²

21. rectangular prism
    l = 7.3 yd
    w = 4.1 yd
    h = 6.5 yd
    208.06 yd²

The map distance is given. Write and solve a proportion to find the actual distance. Use a map scale of 1 cm = 68 mi.

22. 4 cm
    $\frac{1}{68} = \frac{4}{n}$; n = 272 mi

23. $10\frac{1}{2}$ cm
    $\frac{1}{68} = \frac{10\frac{1}{2}}{n}$; n = 714 mi

24. 5 cm
    $\frac{1}{68} = \frac{5}{n}$; n = 340 mi

25. 15 cm
    $\frac{1}{68} = \frac{15}{n}$; n = 1,020 mi

26. $7\frac{1}{2}$ cm
    $\frac{1}{68} = \frac{7\frac{1}{2}}{n}$; n = 510 mi

27. $12\frac{1}{4}$ cm
    $\frac{1}{68} = \frac{12\frac{1}{4}}{n}$; n = 833 mi

Name _____   LESSON 28.3

Number Patterns and Functions

Write an equation to represent the function.

1.
w	3	9	15	21	33
l	1	3	5	7	11

$l = w \div 3$

2.
x	2	4	6	8	10
y	6	10	14	18	22

$y = 2x + 2$

3.
s	14	12	8	6	4
t	12.8	10.8	6.8	4.8	2.8

$t = s - 1.2$

4.
m	2	6	7	9	11
n	8	24	28	36	44

$n = 4m$

Write an equation to represent the function. Then find the missing term.

5.
j	11	15	19	23	27
k	23	27	31	35	39

$k = j + 12$

6.
a	32	26	22	18	14
b	16	13	11	9	7

$b = \frac{a}{2}$

7.
e	2	6	7	9	11
f	10	30	35	45	55

$f = 5e$

8.
x	2	3	4	5	6
y	5	8	11	14	17

$y = 3x - 1$

Write an equation for the function. Possible answers are given.

9. The width of a rectangle is $\frac{1}{3}$ its length. ____$w = \frac{1}{3}l$____

10. An elevator travels at the rate of 5 floors per minute. ____$r = 5m$____

11. Each person on the bus has two suitcases. ____$p = s \div 2$____

Mixed Review

Write as a percent.

12. 0.002   13. $\frac{7}{20}$ 14. 1.18   15. $\frac{1}{25}$

____0.2%____   ____35%____   ____118%____   ____4%____

Write the prime factorization in exponent form.

16. 90   17. 252   18. 675   19. 500

__$2 \times 3^2 \times 5$__   __$2^2 \times 3^2 \times 7$__   __$3^3 \times 5^2$__   __$2^2 \times 5^3$__

PW118 Practice

Name _____

LESSON 28.4

Geometric Patterns

Draw the next three figures in the pattern.

1.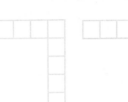

2.

3.

4.

Draw the next two figures in the pattern.

5.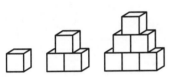

6.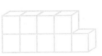

Mixed Review

Tell which measurement is more precise.

7. 38 oz or 6 lb   8. 14.5 gal or 20 c   9. 41 cm or 41 m

   _38 oz_   _20 c_   _41 cm_

Write the fraction as a mixed number or a whole number.

10. $\frac{51}{8}$ _$6\frac{3}{8}$_   11. $\frac{19}{5}$ _$3\frac{4}{5}$_   12. $\frac{23}{2}$ _$11\frac{1}{2}$_   13. $\frac{91}{7}$ _13_

Practice PW119

Name _____

LESSON 29.1

Transformations of Plane Figures

Tell which type or types of transformations the second figure is of the first figure. Write *translation, rotation,* or *reflection.*

1.

   ___rotation___

2.

   ___translation___

3.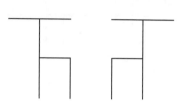

   ___reflection___

Draw a 90° rotation and a horizontal reflection of each original figure.
Check students' work.

4.

5.

Mixed Review

Find the volume.

6. rectangular prism
   $l = 7$ yd; $w = 9$ yd
   $h = 14$ yd

   ___882 yd³___

7. cube
   side = 12 cm

   ___1,728 cm³___

8. rectangular prism
   $l = 3.6$ m; $w = 0.8$ m
   $h = 1.5$ m

   ___4.32 m³___

Use a proportion to convert to the given unit.

9. 10.5 m = __1,050__ cm
10. 670 g = __0.67__ kg
11. 56 L = __56,000__ mL

PW120 Practice

Name _____

LESSON 29.2

Tessellations

Make the tessellation shape described by each pattern. Then form two rows of a tessellation. Check students' tessellations.

1.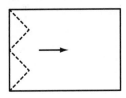
2.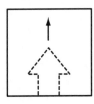
3.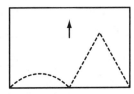
4.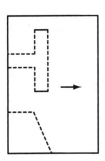
5.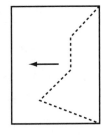
6.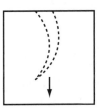

Tell whether the shape can be used to form a tessellation. Write *yes* or *no*.

7.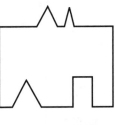

   __no__

8.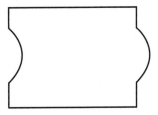

   __yes__

9.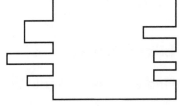

   __no__

Mixed Review

Find the area of each figure.

10. square
    side = 8.3 in.

    __68.89 in.²__

11. triangle
    $b = 16$ cm; $h = 11$ cm

    __88 cm²__

12. rectangle
    $l = 37$ ft; $w = 21$ ft

    __777 ft²__

Write the difference in simplest form.

13. $7\frac{1}{3} - 3\frac{3}{4}$

    $3\frac{7}{12}$

14. $4\frac{2}{5} - 1\frac{2}{3}$

    $2\frac{11}{15}$

15. $9\frac{3}{8} - \frac{4}{5}$

    $8\frac{23}{40}$

16. $30 - 27\frac{7}{8}$

    $2\frac{1}{8}$

Practice   PW121

Name _____

LESSON 29.3

Problem Solving Strategy: Make a Model

Solve the problem by making a model.

1. Carol is making a design from the shape below. She wants the shape to tessellate a plane. Can she use this shape?

   _____yes_____

2. Can this shape be used to form a tessellation? Explain.

   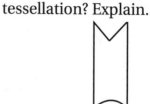

   ___No. The shapes do not fit___
   ___together without gaps.___

3. Adam drew this shape for a tile floor. Will his shape form a tessellation?

   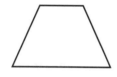

   _____yes_____

4. Draw a figure that does NOT tessellate a plane.

   Check students' drawings.

Mixed Review

Find the unknown dimension.

5. scale: 1 in.:5 yd

   drawing length: 8 in.

   actual length: __40 yd__

6. scale: 3 cm:35 m

   drawing length: __45 cm__

   actual length: 525 m

7. scale: 5 cm:3 mm

   drawing length: 27 cm

   actual length: __16.2mm__

Find the angle measures.

8. ∠1 = __95°__

9. ∠2 = __52°__

10. ∠3 = __33°__

11. ∠4 = __52°__

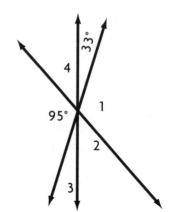

PW122 Practice

Name _____

LESSON 29.4

Transformations of Solid Figures

Tell how many ways you can place the solid figure on the outline.

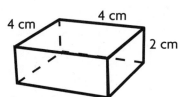

1. [square outline 4 cm × 4 cm]

   _____8 ways_____

2. [rectangle outline 2 cm × 4 cm]

   _____8 ways_____

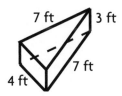

3. [rectangle outline 3 ft × 7 ft]

   _____4 ways_____

4. [triangle outline 4 ft base, 7 ft side]

   _____2 ways_____

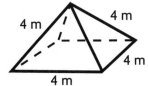

5. [square outline 4 cm × 4 cm]

   _____4 ways_____

6. [triangle outline 4 m sides]

   _____12 ways_____

Mixed Review

Write the next three possible terms in each sequence.

7. 5, 7, 9, 11, 13, _____
   _____15, 17, 19_____

8. 2, 6, 18, 54, _____
   _____162, 486, 1,458_____

9. 12, 8, 4, 0, _____
   _____ ⁻4, ⁻8, ⁻12_____

10. ⁻5, 10, ⁻20, 40, _____
    _____ ⁻80, 160, ⁻320_____

Find the circumference of the circle. Use 3.14 for pi (π).

11. $r = 15.25$ km
    _____95.77 km_____

12. $d = 33$ m
    _____103.62 m_____

13. $r = 5.5$ ft
    _____34.54 ft_____

14. $d = 46$ in.
    _____144.44 in._____

15. $r = 8.5$ cm
    _____53.38 cm_____

16. $d = 12.3$ dm
    _____38.622 dm_____

Practice   PW123

Name _____   LESSON 29.5

Symmetry

Draw the lines of symmetry.

1. 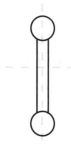   2.    3.

Complete the other half of the figure across the line of symmetry.

4. 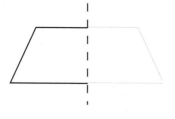   5.    6.

Tell whether each figure has rotational symmetry, and, if so, identify the symmetry as a fraction of a turn and in degrees.

7.    8.    9.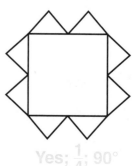

Yes; $\frac{1}{4}$; 90°   Yes; $\frac{1}{2}$; 180°   Yes; $\frac{1}{4}$; 90°

Mixed Review

Write the percent as a decimal.

10. 26%   11. 9%   12. 71%   13. 16.5%

    0.26   0.09   0.71   0.165

Write the mixed number as a fraction.

14. $3\frac{3}{4}$ 15. $7\frac{1}{8}$ 16. $1\frac{5}{6}$ 17. $10\frac{3}{5}$

    $\frac{15}{4}$ $\frac{57}{8}$ $\frac{11}{6}$ $\frac{53}{5}$

PW124 Practice

LESSON 30.1

Name _____

Inequalities on a Number Line

Graph the solutions of the inequality.

1. $x > 5$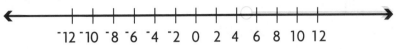

2. $x \leq {}^-2$

3. $x \geq {}^-1$

4. $x < 7$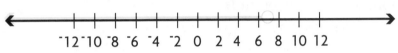

Solve the inequality and graph the solutions.

5. $x + 3 > 7$ $x > 4$

6. $n - 5 < 3$ $n < 8$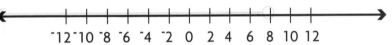

7. $2p \leq 6$ $p \leq 3$

8. $k + 7 > 7$ $k > 0$

For 9–10, write an algebraic inequality for the word sentence.

9. The value of m is greater than or equal to fifteen.

   $m \geq 15$

10. The value of w is less than negative forty-three.

    $w < {}^-43$

Mixed Review

Find the area of each figure.

11. a parallelogram with base 6.2 ft and height 2.7 ft

    16.74 ft^2

12. a trapezoid with bases 23 cm and 19 cm and height 11.4 cm

    239.4 cm^2

13. a parallelogram with base 32.8 m and height 8.4 m

    275.52 m^2

Use a proportion to change to the given unit.

14. 3,500 lb = $1\frac{3}{4}$ T

15. 126 in. = $3\frac{1}{2}$ yd

16. 41 qt = $10\frac{1}{4}$ gal

Practice   PW125

Name _____

LESSON 30.2

Graph on the Coordinate Plane

Write the ordered pair for each point on the coordinate plane.

1. point A
2. point B
3. point C

   (1,3)   (⁻4,⁻2)   (3,⁻2)

4. point D
5. point E
6. point F

   (⁻2,5)   (0,3)   (⁻4,3)

7. point G
8. point H
9. point J

   (5,0)   (⁻6,⁻4)   (2,⁻5)

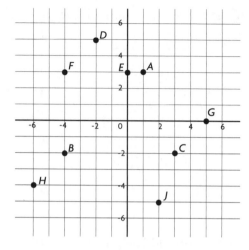

Use the coordinate plane above. Identify the points located in the given quadrant.

10. I   A
11. II   D, F
12. III   B, H
13. IV   C, J

Plot the points on the coordinate plane.   Check students' graphs.

14. S (0,5)
15. T (2,2)
16. U (⁻5,4)
17. V (⁻2,⁻2)
18. W (5,⁻2)
19. X (6,0)

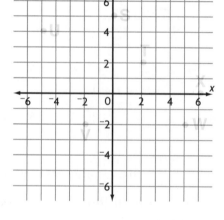

Mixed Review

Find the percent of the number.

20. 78% of 152
21. 12% of 37
22. 57% of 238
23. 0.6% of 200

    118.56   4.44   135.66   1.2

Find the circumference of each circle. Use 3.14 for π.

24. $r = 7$ in
25. $d = 12$ ft
26. $d = 15$ m
27. $r = 30$ cm

    43.96 in   37.68 ft   47.1 m   188.4 cm

PW126 Practice

Name _____

Graph Functions

LESSON 30.3

Complete the function table.

1.
x	1	2	3	4	5
y	3	4	5	6	7

2.
x	1	2	3	4	5
y	5	10	15	20	25

3. Graph the data from Exercise 1 on the coordinate plane.

4. Graph the data from Exercise 2 on the coordinate plane.

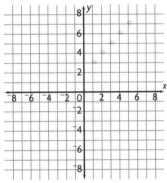
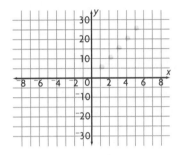

5. Write an equation relating y to x for the data in Exercise 1.

   $y = x + 2$

6. Write an equation relating y to x for the data in Exercise 2.

   $y = 5x$

7. Use the equation $y = x - 5$ to make a function table. Use the integers from $^-3$ to 3 as values of x.

x	$^-3$	$^-2$	$^-1$	0	1	2	3
y	$^-8$	$^-7$	$^-6$	$^-5$	$^-4$	$^-3$	$^-2$

Mixed Review

Find the number of possible choices for each situation.

8. 5 flavors of ice cream and 4 toppings

   20 choices

9. 4 shirts, 6 ties, and 2 jackets

   48 choices

10. 3 kinds of pancakes and 4 kinds of syrup

    12 choices

Evaluate the expression for $n = ^-3, ^-1,$ and 4.

11. $3n - 2(n + 5) + n$

    $^-16, ^-12, ^-2$

12. $5 + 2n - (6 + n)$

    $^-4, ^-2, 3$

Practice **PW127**

Name _____

LESSON 30.4

Problem Solving Skill: Make Generalizations

Solve by making a generalization.

Anita uses 2.5 c of flour to make a dozen muffins. The table shows the number of dozens of muffins made, x, for different amounts of flour, y.

x (doz)	2	4	6	8	10
y (c)	5	10	15	20	25

1. What equation can be used to show the amount of flour that Anita uses? **D**

   A $y = x - 2.5$ C $y = x + 2.5$
   B $y = x \div 2.5$ D $y = 2.5x$

2. How much flour does Anita use to make 16 dozen muffins? **J**

   F 6.4 c   H 20 c
   G 18.5 c   J 40 c

Anita charges $0.75 for each muffin.

3. Write an equation to show the cost, m, when Anita sells n muffins.

   _____ $m = \$0.75n$ _____

4. How much will Anita charge for 15 muffins?

   _____ $11.25 _____

Rick spends $8 on supplies for his dog-grooming business. The table shows his profit, y, for several income amounts, x.

x	$30	$35	$40	$45
y	$22	$27	$32	$37

5. What equation can be used to show Rick's profit? **B**

   A $y = 8 - x$ C $y = \frac{1}{8}x$
   B $y = x - 8$ D $y = x + 8$

6. How much profit did Rick make if he earned $105? **J**

   F $113   H $101
   G $109   J $97

Rick charges $35 for a regular dog grooming.

7. What equation can Rick use to show the amount that he earns, y, when he grooms x dogs?

   _____ $y = 35x$ _____

8. How much will Rick earn if he grooms 12 dogs?

   _____ $420 _____

Mixed Review

Find the simple interest.

9. principal: $2,200
   rate: 7.3%
   time: 4 yr

   _____ $642.40 _____

10. principal: $14,000
    rate: 6.7%
    time: 8 yr

    _____ $7,504 _____

11. principal: $35,000
    rate: 8.2%
    time: 12 yr

    _____ $34,440 _____

PW128 Practice

Name _____

LESSON 30.6

Graph Transformations

Transform the figure according to the directions given.
Name the new coordinates.

1.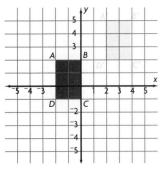

Translate 4 units right and 3 units up.

$A'(2,5), B'(4,5),$
$C'(4,2), D'(2,2)$

2.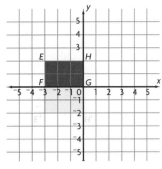

Reflect across the x-axis.

$E'(^-3,^-2), F'(^-3,0),$
$G'(0,0), H'(0,^-2)$

3.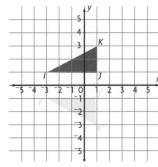

Reflect across the x-axis.

$I'(^-3,^-1), J'(1,^-1),$
$K'(1,^-3)$

Rotate the figure around the origin according to the directions given.
Name the new coordinates.

4. 90° clockwise

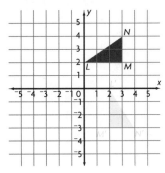

$L'(2,0), M'(2,^-3),$
$N'(4,^-3)$

5. 180° counterclockwise

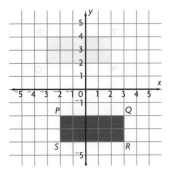

$P'(2,2), Q'(^-3,2),$
$R'(^-3,4), S'(2,4)$

6. 90° counterclockwise

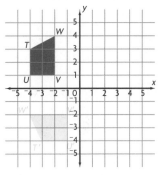

$U'(^-1,^-4), V'(^-1,^-2),$
$W'(^-4,^-2), T'(^-3,^-4)$

Mixed Review

Find the next three possible terms in each sequence.

7. 4, 12, 36, 108, . . .

324; 972; 2,916

8. 27, 19, 11, 3, . . .

$^-5, ^-13, ^-21$

9. 7, 11, 18, 29, . . .

47, 76, 123

Solve and check.

10. $h + 12 = 37$

$h = 25$

11. $m + 8 = 19$

$m = 11$

12. $^-43 = 4 + p$

$p = ^-47$

Practice PW129